WILD BERRIES
of ONTARIO

Fiona Hamersley Chambers
Cory Harris

with contributions from
Andy MacKinnon, Linda Kershaw
John Thor Arnason, Patrick Owen & Amanda Karst

Lone Pine Publishing

The Publisher: Lone Pine Publishing
2311 – 96 Street
Edmonton, Alberta T6N 1G3
Website: www.lonepinepublishing.com

Library and Archives Canada Cataloguing in Publication

Hamersley Chambers, Fiona, 1970–
    Wild berries of Ontario / Fiona Hamersley Chambers,
Cory Harris; with contributions from Andy MacKinnon ... [et al.]

Includes bibliographical references and index.
ISBN 978-1-55105-867-2

    1. Berries—Ontario—Identification.
I. Harris, Cory, 1977– II. Title.

QK203.O5H34 2012   581.4'6409713   C2011-908414-7

Editorial Director: Nancy Foulds
Project Editor: Gary Whyte
Editorial Support: Kathy van Denderen, Kelsey Attard
Production Manager: Gene Longson
Book Design: Lisa Morley
Layout & Production: Alesha Braitenbach
Cover Design: Gerry Dotto

Map Source: Commission for Environmental Cooperation. 2009. Ecological Regions of North America: Level III. CEC. Montreal, Canada.

We acknowledge the financial support of the Government of Canada through the Canada Book Fund (CBF) for our publishing activities.

PC: 16

Dedication

To three incredible women who I am fortunate to have in my life: my mother Sarah Richardson for being so wonderfully enthusiastic about plants and for doing lots of quality child care while I write, my second mum Vicky Husband for showing me what you can do if you don't take "no" for an answer and persist against the odds, and to Nancy Turner for showing me what you can accomplish by always being positive and kind. And most of all to my boys Hayden and Ben. You guys are the best foragers and berry testers I know and you give me such joy.

–FHC

To my belle-mère Louise, whose enthusiasm, patience and perseverance (in berry picking and in life) is inspiring, to my wonderful daughters Émilie and Cléa, and to all the men, women and children working hard to ensure we all have wild berries to pick for generations to come.

–CH

Acknowledgements

The following people are thanked for their valued contribution to this book:

- ❧ The many photographers who allowed us to use their photographs;

- ❧ Lone Pine's editorial and production staff;

- ❧ And most importantly, the indigenous peoples, settlers, botanists and writers who kept written records or oral accounts of the many uses of the berries featured in this book. We are so grateful to our family, friends and teachers who have shared their knowledge and enthusiasm about our native plants and their uses.

Contents

Dedication and Acknowledgements 3

Plants at a Glance 6

Map — Ontario Ecological Regions 10

Introduction 11

What is a "Berry"? 13 • Edibility Scale 18 • A Few Gathering Tips 21 • A Cautionary Note 25 • Disclaimer 30

The Berries

Trees & Shrubs 32

Apple 32 • Junipers 36 • Hawthorns 42 • Mountain Ashes 48 • Wild Roses 50 • Black Chokeberry 54 • Cherries 56 • Chokecherry 60 • Sumacs 64 • Blackberries 66 • Raspberries 70 • Dewberries 74 • Cloudberry 76 • Thimbleberry 78 • Mulberries 80 • Barberries 82 • Tall Oregon-grape 84 • Sassafras 86 • Currants 88 • Gooseberries 92 • Prickly Currant 96 • Serviceberries 98 • Dogwoods 102 • Huckleberries 106 • Blueberries 110 • Cranberries 116 • False Wintergreens 120 • Bearberries 122 • Black Crowberry 124 • Riverbank Grape 126 • Elderberries 128 • Bush-cranberries 132 • Wayfaring Tree 138 • Soapberry 140 • Common Hackberry 142 • Silverberry 144

Wildflowers 146

Partridge Berry 146 • Yellow Clintonia 148 • Twisted-stalks 150 • Canada Mayflower 154 • False Solomon's-seals 156 • Rough-fruited Fairybells 160 • Indian Cucumber Root 162 • Solomon's-seals 164 • Strawberries 166 • Ginseng 170 • Sarsaparillas 172 • Bunchberry 174 • Ground Cherries 176 • Northern Comandra 178

Poisonous Plants 180

Winterberry 180 • Mountain Holly 182 • Buckthorns 184 • Canadian Yew 186 • Poison-ivy & Poison-oak 188 • Poison Sumac 190 • Devil's Club 192 • Honeysuckles 194 • Snowberries 198 • Baneberries 202 • Nightshades 204 • Jack-in-the-pulpit 206 • Pokeweed 208 • Blue Cohoshes 210 • May-apple 212 • Leatherwood 214 • False Virginia Creeper 216 • Canada Moonseed 218

Glossary 220

References 224

Index 226

Photo Credits 231

About the Authors 232

LIST OF RECIPES

Dried Fruit. 22
huckleberries, strawberries, thimbleberries, blueberries, saskatoons, cranberries or currants

Frozen Wild Fruit 22
wild berries

Rosehip Jelly 53
rosehips, apples

Indian Lemonade 65
sumac flower spikes, frozen blueberries

Blackberry Syrup 69
blackberries

Berry Blackberry Cordial 69
blackberries or other juicy berries (raspberries, thimbleberries or berry combination)

Wild Berry Dressing. 73
mixed tangy wild berries (raspberries, thimbleberries or blackberries)

Berry Fruit Leather 77
crushed berries (one kind or a mix)

Wild Berry Juice. 79
sweet berries (blueberries, thimbleberries or blackberries)

Blackberry and Oregon-grape Jelly 85
blackberries, Oregon-grape berries

Saskatoon Squares 99
saskatoons or blueberries

Pemmican 99
saskatoons or blueberries

Tom's Huckleberry Pie 108
red or black huckleberries

Blueberry Cobbler .115
blueberries or huckleberries

Fruit Popsicles 115
wild berries

Cranberry Chicken .119
cranberries

Riverbank Grape Jelly. 127
riverbank grapes

Indian Ice Cream 141
soapberries

Wild Berry Muffins .169
mixed wild berries (strawberries, thimbleberries, blueberries, huckleberries, dewberries, etc.)

Plants at a Glance

TREES & SHRUBS

Apple p. 32

Junipers p. 36

Hawthorns p. 42

Mountain Ashes p. 48

Wild Roses p. 50

Black Chokeberry p. 54

Cherries p. 56

Chokecherry p. 60

Sumacs p. 64

Blackberries p. 66

Raspberries p. 70

Dewberries p. 74

Cloudberry p. 76

Thimbleberry p. 78

Mulberries p. 80

Barberries p. 82

Tall Oregon-grape p. 84

Sassafras p. 86

Currants p. 88

Gooseberries p. 92

Prickly Currant p. 96

Serviceberries p. 98

Dogwoods p. 102

Huckleberries p. 106

Blueberries p. 110

Cranberries p. 116

False Wintergreens p. 120

Bearberries p. 122

Black Crowberry p. 124

Riverbank Grape p. 126

Elderberries p. 128

Bush-cranberries p. 132

Wayfaring Tree p. 138

Soapberry p. 140

Common Hackberry p. 142

Silverberry p. 144

WILDFLOWERS

Partridge Berry p. 146

Yellow Clintonia p. 148

Twisted-stalks p. 150

False Solomon's-seals p. 156

Rough-fruited Fairybells
p. 160

Canada Mayflower p. 154

Indian Cucumber Root
p. 162

Solomon's-seals p. 164

Strawberries p. 166

Ginseng p. 170

Sarsaparillas p. 172

Bunchberry p. 174

Ground Cherries p. 176

Northern Comandra
p. 178

POISONOUS PLANTS

Winterberry p. 180

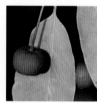

Mountain Holly p. 182

Buckthorns p. 184

Canadian Yew p. 186

Poison-ivy & Poison-oak p. 188

Poison Sumac p. 190

Devil's Club p. 192

Honeysuckles p. 194

Snowberries p. 198

Baneberries p. 202

Nightshades p. 204

Pokeweed p. 208

Blue Cohoshes p. 210

Jack-in-the-pulpit p. 206

May-apple p. 212

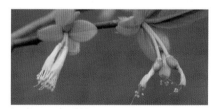

Leatherwood p. 214

False Virginia Creeper
p. 216

Canada Moonseed
p. 218

Ontario Ecological Regions

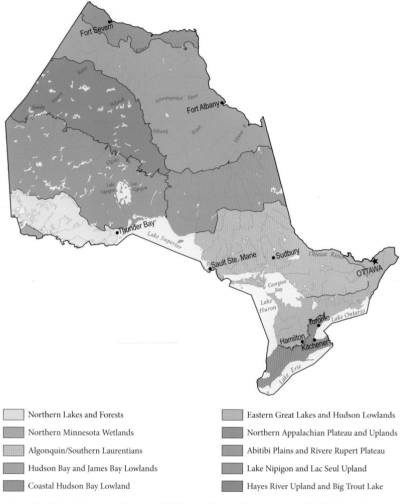

Northern Lakes and Forests	Eastern Great Lakes and Hudson Lowlands
Northern Minnesota Wetlands	Northern Appalachian Plateau and Uplands
Algonquin/Southern Laurentians	Abitibi Plains and Rivere Rupert Plateau
Hudson Bay and James Bay Lowlands	Lake Nipigon and Lac Seul Upland
Coastal Hudson Bay Lowland	Hayes River Upland and Big Trout Lake

Source: Commission for Environmental Cooperation. 2009. Ecological Regions of North America: Level III. CEC. Montreal, Canada.

Introduction

Woodland strawberry (*Fragaria vesca*)

It's difficult to find someone who does not enjoy eating a berry. Juicy, sweet, tart, sometimes sour, generally bursting with flavour and very good for you—wild berries are gifts from the land, treasures to be discovered on a casual hike or potentially a lifesaving food if you're unfortunate enough to get lost in the woods. Berries have a long history of human use and enjoyment as food and medicine, in ceremonies and for ornamental and wildlife value. Our ancestors needed to know as a matter of survival which berries were edible or poisonous, where they grew and in which seasons and how to preserve them for non-seasonal use. These early peoples often went to great lengths to manage their wild berry resources: pruning, coppicing, burning, transplanting and even selectively breeding some wild species into the domesticated ancestors of many of our modern fruit varieties.

Today, many of us live in urban environments where the food on our plate and in our pantries comes from distant places. The first strawberry is no longer an eagerly awaited and delectable harbinger of the summer to come. Rather than a fleetingly sweet June moment, these fruits are now available on our grocery store shelves almost year-round. A sad result of this convenience and lack of seasonality is that this store-bought fruit has little resemblance to its forebears. Grocery store strawberries, for example, are generally not properly ripe, don't taste of much and are not loaded with nutrients. As we become more and more disconnected from our food sources, it is even possible that we are forgetting what a "real" berry tastes like. Perhaps part of the exceptional taste of wild fruit is the thrill of the hunt and the discovery of a gleaming berry treasure hanging—sometimes in great profusion—from a vine, bush or herb. These wild berries are only available for a short time during the year and we must increasingly travel to find them growing in their native state. We must

Wild red raspberry (*Rubus idaeus*)

make an effort to discover them in the wild or find a reputable source for those plants that will grow in our home gardens.

The wild berries described in this book are, for the most part, not available in stores. When they are, they are often very expensive. A berry in its prime state of ripeness is juicy and delicate, and therefore does not travel well. What a pity, as slowly savouring one of these fruits at their peak of perfection plucked fresh off the plant is one of the great joys in life. What better way to spend a warm summer's day than wandering hillsides, country roads or forest edges with friends and family in search of these delectable morsels? Wild berry gathering builds community and family and is a great way to connect you and your children to nature. In winter, a spoonful of these frozen or preserved wild fruits will bring back the taste of summer for a delicious moment. Our young children, Fiona's two boys and Cory's two girls, love to join in berry gathering and are proud when they share the fruits of their labour, whether eaten fresh or in jams and baking that they helped to make. It is our hope that, whether you are a seasoned gatherer or a new enthusiast, this book will help guide you to experience and share in this wonderful and generous gift of nature.

Why Learn to Identify and Gather Wild Berries?

Berries gathered in the wild generally have superb flavour and can be gathered when they are properly ripe.

These fruits are not only delicious, but contain important nutrients and phytochemicals (compounds produced by plants, many of which promote good health) that are increasingly lacking from our commercially available fruits. Many wild berries are high in vitamin C and also contain carbohydrates, proteins and important nutrients such as vitamin A, thiamine, calcium, iron and other trace minerals. While most people will obtain this guidebook in order to enjoy wild berries on hikes and outings, the information that you learn here could also save your life if you ever get stranded or lost in the backcountry. Be warned, though! Gathering wild berries can also be considered a dangerous "gateway" into the more complex realm of preserving and cooking with these fruits as well as growing them in your own backyard. Once you start on this journey it can become rather addictive and even spread to friends and family!

What Is Not Covered in this Guide?

Although this book should enable you to identify most native berry species in your region, it is not intended as a complete reference guide. A section on references and further reading is provided for those wishing to study these plants in greater detail. Some berry species are so rarely found or have such a restricted range that it would not be useful to include them here. There are many excellent resources already available to help you understand the cultivation and use of

domesticated fruit species, so these are also not covered. Nuts and seeds are not considered "berries" in the common sense of the word, so these are excluded. The same goes for most cones, with the exception of the modified, berry-like cones of junipers, Canadian yew and Eastern red-cedar, which we could not resist including in this book.

What Is a "Berry"?

In this guide, "berry" is used in the popular sense of the word, rather than in strictly botanical terms, and includes any small fleshy fruit. Technically, a "berry" is a fleshy, simple fruit produced from a single ovary that contains one or more ovule-bearing structures (carpels) that each contain one or more seeds. The outside covering (the endocarp) of a berry is generally soft, moist and fleshy, most often in a globular shape. Roughly translated, a berry is really a seed(s) packaged in a tasty moist pulp that encourages animals to eat the fruit and distribute the seeds far and wide from the parent plant so that these offspring can grow and flourish. "True berries" include currants, huckleberries, blueberries and grapes.

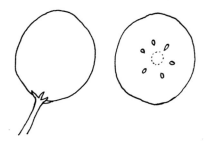

Velvetleaf blueberry (*Vaccinium myrtilloides*)

Botanically, however, what we call a berry often includes simple fleshy fruits such as drupes and pomes. The botanical definitions of different types of fruit are provided below for general interest and are also sometimes mentioned, where appropriate, in the text.

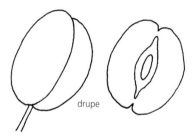

drupe

A "drupe" is a fleshy stone fruit that closely resembles a berry but has a single seed or stone with a hard inner ovary wall that is surrounded by a fleshy tissue. Wild fruit in this category include highbush cranberries and bunchberries; some domestic fruit examples are cherries or plums.

A "compound drupe" or "aggregate" fruit ripens from a flower that has multiple pistils, all of which ripen together into a mass of multiple fruits,

called "drupelets." A drupelet is a collection of tiny fruit that forms within the same flower from individual ovaries. As a result, these fruit are often crunchy and seedy. Wild examples include raspberries and blackberries. Cultivated examples include loganberries, boysenberries and tayberries.

A "multiple fruit" is similar to an aggregate, but differs in that it ripens from a number of separate flowers that grow closely together, each with its own pistil (as opposed to from a single flower with many pistils). Mulberry is the only native Canadian example of a multiple fruit. Tropical examples include pineapples and figs.

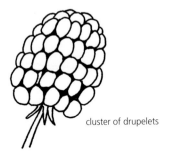

cluster of drupelets

An "accessory fruit" is a simple fruit with some of its flesh deriving from a part other than the ripened ovary. In other words, a source other than the ovary generates the edible part of the fruit. Other names for this type of fruit include pseudocarp, false fruit or spurious fruit. A "pome" is a sort of accessory fruit because it has a fleshy outer layer surrounding a core of seeds enclosed with bony or cartilage-like membranes (it is this inner core that is considered the "true" fruit). Crabapples, serviceberries and hawthorns are wild examples of pomes; apples and pear varieties are domestic examples. Another type of accessory fruit is the strawberry; the main part of the fruit

derives from the receptacle (the fleshy part that stays on the plant when you pick a raspberry) rather than from the ovary. Wintergreen is also an example of an accessory fruit type as the fruit is really a dry capsule surrounded by a fleshy calyx.

A "cone" is a fruit that is made up of scales (sporophylls) that are arranged in a spiral or overlapping pattern around a central core, and in which the seeds develop between the scales. In some cone-producing species such as juniper and yew, the fleshy modified scales fuse to form a berry-like structure.

cone

A "hip" has a collection of bony seeds (achenes), each of which comes from a single pistil, covered by a fleshy receptacle that is contracted at the mouth. The rose hip (which is also an accessory fruit) is our only example of a hip.

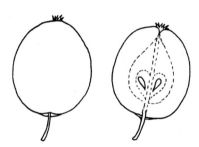

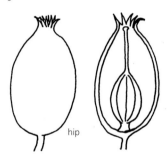

hip

Common hawthorn (*Crataegus monogyna*)

The Species Accounts

In this book, species are organized by growth form into three main sections—Trees and Shrubs, Flowering Plants, and Poisonous Plants. Closely related or similar plants are grouped together for comparison and the section on poisonous plants is conveniently located at the end of the book.

This book includes plants common in Ontario that have been used by people from ancient times to the present. Each account has a detailed description for each plant including plant form, leaf structure, habitat and range, as well as fruit form, colour, season and—of course—edibility. This description, in addition to colour photographs and illustrations, will help you ensure safe plant and berry identification. Information on traditional and contemporary uses for food, medicines and material culture are also included for general interest.

The information in each species account is presented in an easy-to-follow format. In addition to the opening general discussion, each account includes subheadings of Edibility (see below for edibility scale), Fruit (a description of the look and taste of the fruit), Season (flowering and fruiting seasons) and Description (a detailed description of the plant, flowers, habitat and range).

Many species accounts focus on a single species, but if several similar species have been used in the same ways, two or more species may be described together. In these "group" accounts, you will find a general description for the group, but there will also be separate paragraphs in which the individual species in the group are described in specific detail. So, for example, the Hawthorns (*Crataegus* spp.) account describes historical and modern uses in the opening discussion, then includes subheadings of Edibility, Fruit, Season, and Description for all hawthorns. These subheadings are followed by separate paragraphs with important

specific information (including identification and location details) for each of the numerous hawthorns in the province (black hawthorn, cockspur hawthorn, common hawthorn, dotted hawthorn, downy hawthorn, fanleaf hawthorn, fireberry hawthorn, fleshy hawthorn, scarlet hawthorn).

Where appropriate, you will find an "Also called" subheading that describes other common and scientific names for each species. These "Also called" names are found below the main account title and at the end of individual species description paragraphs inside the account.

Cockspur hawthorn (*Crataegus crus-galli*)

Edibility

All accounts contain a useful scale of edibility for each species. Although we have a wonderful variety of native berries, it is useful to know which ones are worth our time pursuing; which ones, although considered "edible," are better left for the birds or as a famine food; and which ones are toxic or poisonous.

Northern dewberry (*Rubus flagellaris*)

Highly edible describes those berries that are most delicious and are well worth gathering and consuming. A wild strawberry or serviceberry is considered highly edible. An example of such a fruit is the northern dewberry.

Bunchberry (*Cornus canadensis*)

Edible describes those berries that are still tasty, but not as good as the prime edible species. An example of such a fruit is bunchberry.

Silverberry (*Elaeagnus commutata*)

Not palatable describes berries you can eat without any ill effects but are perhaps not worth the effort to harvest given their lack of flavour, their bitterness, relatively large seeds or lack of fleshiness. It is useful to know about these species in case you are desperate to snack on something in the woods, but they are not berries that you would actively gather to make a pie! Silverberry is an example of such a species.

Common juniper (*Juniperus communis*)

Edible with caution (toxic) are berries that are palatable, but have differing reports as to their edibility, or perhaps they are only toxic if you eat large amounts or if they haven't been prepared properly or are unripe. Berries of our native juniper species are an example under this category.

Red baneberry (*Actaea rubra*)

Poisonous berries are ones that are definitely poisonous and should not be eaten. Examples of poisonous berries are red and white baneberries.

Species Range and Season

The species ranges and habitats described in this guide were obtained from *Shrubs of Ontario, The ROM Field Guide to Wildflowers of Ontario, Flora of North America*, the United States Department of Agriculture Plants Database website, regional field guides, personal experience and interviews, academic papers and other sources. Despite all due diligence being taken, however, this description is not universal or foolproof. Plants sometimes either grow outside of their reported ranges, or cannot be found within the described habitat. Likewise, the season given for flowering and fruit production for each species is an average. Specific microclimates like deep valley bottoms or high mountaintops will necessarily produce a wide range of flowering and fruiting variability for the same plant. Berry plants also produce fruit of differing quality and quantity from year to year, depending on factors such as plant age and health, changes in temperature and moisture, or insect infestation. Some berries, such as rosehips, are best harvested later in the year after the first frost sweetens the fruit.

European mountain ash (*Sorbus aucuparia*)

Description

The plant description and the accompanying photos and illustrations are important parts of each species account. Each plant description begins with a general outline of the form of the species or genus named at the top of the page. Detailed information about diagnostic features of the leaves, flowers and fruits is then provided. Flowering time is included as part of the flower description to give some idea of when to look for blooms, and a general fruiting season is also included. If two or more species of the same genus have been used for similar purposes, several of the most common species may be illustrated and their distinguishing features described.

Plant characteristics such as size, shape, fruiting, colour and hairiness vary with season and habitat and with the genetic variability of each species. Identification can be especially tricky when plants have not yet flowered or fruited. If you are familiar with a species and know its leaves or roots at a glance, you may be able to identify it at any time of year (from very young shoots to the dried remains of last year's plants), but sometimes a positive ID is just not possible.

General habitat information is provided for each species to give you some idea of where to look for a plant. The habitat description provides information about general habitat (e.g., in moist, mossy forests), elevation (e.g., low to montane elevations) and range (e.g., from the northern part of a province to its southern regions). However, the habitat information included for each species is meant as a general guide only; plants often grow in a variety of habitats over a broad geographical range.

The origin of non-North American species is also noted. The flora of many areas has changed dramatically over the past 200 years, especially in and around human settlements. European settlers brought many plants with them, either accidentally (in ship ballast, packing and livestock bedding) or purposely (for food, medicine, fibre, ornamental value, etc.). Some of these introduced species produce fruit and have thrived, and some are now considered weeds on disturbed sites across much of Canada. An example of such an introduced species is European mountain ash.

What's in a Name?

Both common and scientific names are included for each plant and are based on authoritative sources including *Flora of North America, Flora of Canada* (Scoggan 1978–1979), the

A Few Gathering Tips

1. Gather only species that are common and abundant, and never take all the fruit off one plant. Even then, a cautious personal quota will still deplete the plants if too many people gather them in one area. Remember, plants growing in harsh environments (e.g., northern areas, alpine, desert) might not have enough energy to produce flowers and fruits every year. Also, don't forget the local wildlife. Survival of many animals can depend on access to the fruits that you are harvesting.

2. Never gather from plants that grow in protected and/or heavily used areas such as parks and nature preserves. Doing so is not only wrong, but is also often illegal. Be sure to check the regulations for the area you are visiting.

3. Take only what you need, and damage the plant as little as possible. If you want to grow a plant in your garden, try propagating it from seed or a small cutting rather than transplanting it from the wild.

4. Don't take more than you will use. If you are gathering a plant for food, taste a sample. You may not like the taste of the berries, or the fruit at this site may not be as sweet and juicy as the ones you gathered last year.

5. Gather berries only when you are certain of their identity. Many irritating and poisonous plants grow wild in Canada, and some of them resemble edible or medicinal species. If you are not positive that you have the right plant, don't use it. It is better to eat nothing at all than to be poisoned!

Recipes

Simple recipes for cooking, preserving or enjoying berries fresh off the plant are included in this book. Even the most amateur berry gatherer or cook can produce some delicious results to enjoy with friends and family both in the heat of summer and later during the long winter months. Although the recipes call for specific berries, you can experiment by substituting other fruit. For example, you could try replacing blueberries with serviceberries, cranberries or huckleberries. Be bold and adventurous—it's difficult to make something with fresh berries that tastes bad!

Dried Fruit

It's hard to beat the flavour of home-dried wild berries. Enjoy these special treats out of the bag or add them to your favourite recipes in place of the usual commercial raisins, dried cranberries or blueberries.

A note on berries that dry well: huckleberries, strawberries, thimbleberries (which are a bit crunchy but have a fabulous flavour), blueberries, saskatoons, cranberries and currants. Berries that do not dry as well: seedier fruit such as blackberries or very juicy fruit such as salmonberries; it is better to mash these types of berries either alone or combined with other fruit and make them into fruit leather. Some fruit, such as elderberries, should be cooked before drying to neutralize the toxins present in the fresh fruit.

If some of the berries are much larger than others, cut them in half. All the berries on a tray should be roughly the same size to ensure even drying. Carefully pick through the fruit to remove insects and debris. Do not wash the berries—it will cause them to go mushy. Lightly grease a rimmed baking sheet and spread the berries on the sheet so that they do not touch each other. Place in a food dehydrator or dry in an oven at 140° F overnight, leaving the oven door ajar to allow moisture to escape. Cool and store in an airtight container or Ziploc® bag.

Frozen Wild Fruit the Easy Way

Freezing is the quickest and easiest way to preserve wild berries, making a wonderful snack any time of the year. Choose the best and ripest fruit and carefully remove all unwanted debris and insects. Some fruit, such as elderberries, should be cooked first to neutralize any toxins. Give dusty berries a quick rinse, though the extra water and handling may bruise the fruits and stick them together during freezing.

Most instructions tell you to freeze berries individually on rimmed baking sheets before packing them in Ziploc® bags. However, I have frozen berries very successfully for years in used milk cartons. Open the carton (1 or 2 L size) fully and wash well in warm, soapy water. Allow to air dry. Cartons with the plastic lid and spout do not work for freezing. I collect cartons during the year, wash them, then store them away for when I need lots of them in the summer. Unless a fruit is particularly mushy (like a very ripe wild raspberry), I simply pick through the fruit to clean it, then gently pour the berries into the carton being careful not to let them pack too hard or crush. Push the top of the carton back together the way it was before opening, then firmly push the top edge so that it folds over flat and indents slightly so that it stays shut. Presto, a sealed container that will never get freezer burn, that is easy to label on the top with a marker pen, and that stacks beautifully in the deep freeze!

To get the frozen fruit out, gently squeeze the carton sides to separate the berries, making it easy to pour out the desired quantity before resealing the carton and returning it to the freezer. If the berries are more firmly attached, simply place the carton on the floor and gently stand on it, turning the sides a few times. As a last resort, peel the carton down to the desired level, cut off the exposed fruit chunk with a sharp knife, and put the remainder of the carton in a Ziploc® bag before replacing it in the freezer. I've successfully re-used the same cartons for many years as long as the fruit inside was not too mushy or difficult to extract.

Black huckleberry (*Vaccinium membranaceum*)

Database of Vascular Plants of Canada (Vascan) and Flora Ontario—Integrated Botanical Information System (FOIBIS).

Common names are often confusing. Sometimes, the same common name can refer to a number of different, even unrelated, species. And, at the same time, one common name can even refer to a plant that is edible and to a completely different and unrelated species that is poisonous! For this reason, the scientific name is included for each plant entry.

The two-part scientific name used by scientists to identify individual plants may look confusing, but it is a simple and universal system that is worth taking a few moments to learn about.

Swedish botanist Carolus Linnaeus, who lived from 1707 to 1778, first suggested a system for grouping organisms into hierarchical categories and it is still essentially the same today, almost 300 years after he first developed it! His system differed from other contemporary ones in that it used an organism's morphology (its form and structure) to categorize a species, with a particular emphasis on the reproductive parts, which we now know are the most ancient part of any plant. Another significant benefit of this hierarchical system is that it groups plants into families so that we can better understand and see how they are related to each other. For example, oval-leaved blueberry (*Vaccinium ovalifolium)*, black huckleberry

Common bearberry (*Arctostaphylos uva-ursi*)

(*Vaccinium membranaceum*) and common bearberry (*Arctostaphylos uva-ursi*) are related cousins in the heath (Ericaceae) family. At a more distant level, Linnaeus' system shows us that roses are botanically related to apples. Since the names of organisms in Linnaeus' system follow a standard format and are in Latin (or a Latinized name formed from other words), they are the same in every language around the world, making this a truly universal classification and naming protocol.

In Linnaeus' system, the species name (the "scientific name") has two parts: (1) the genus, and; (2) a species identifier (or specific epithet), which is often a descriptive word. The first part of the scientific name, the genus, groups species together that have common characteristics. The genus name is always capitalized and both parts of the scientific name are either written in *italics* or <u>underlined</u>. The second part, the specific epithet, which is not capitalized, often describes a physical or other characteristic of the organism, honours a person, or suggests something about the geographic range of the species. For example, in the scientific name for bunchberry, *Cornus canadensis*, the specific epithet roughly translates as "from Canada." This apt name describes a species that has a wide distribution across our entire country.

It is important to note, however, that botanists do not always agree on how some plants fit into this system. As a result, scientific names can change over time, or there can sometimes be more than one accepted scientific name for a plant. While this is somewhat annoying and may seem redundant, the important thing to remember is that one scientific name will **never** refer to more than one plant. Thus, if you have identified a wild berry as edible and know its scientific name(s), you can confirm that it is indeed edible and not have to worry that this name may refer to another (possibly deadly poisonous!) plant.

Botanists also sometimes further split species into subsets known as subspecies and varieties. For example, peaches and nectarines are two slightly different varieties of the peach tree, *Prunus persica*. All cultivated apples are known by the Latin *Malus domestica*. If you purchase a golden delicious apple tree at your local plant nursery, the tag should read "*Malus domestica*, variety Golden Delicious."

Giving Back to the Plants

While many of our native wild berries grow in profusion, others are threatened by habitat destruction, overharvesting or climate change. In some areas, such as national parks, harvesting is prohibited. Please do not dig up plants from the wild. Most berry species propagate easily from seed or cuttings, and you can also purchase healthy and responsibly produced plants from reputable nurseries. When you harvest native berries in the wild, it's nice to say "thank you" to the plant by weeding back competing species around its base, spreading some of its seeds in similar habitat a short distance from the parent plant, or appropriately pruning the plant if you know the right technique. There is a long history of humans looking after the plants that support us; taking a few moments to continue this tradition and to teach it to our children is time well spent. By learning about our native berry species and harvesting them, we get to know and respect these plants and may even be moved to help protect and propagate them.

A Cautionary Note

If you cannot correctly identify a plant, you should not use it. Identification is more critical with some plants than with others. For example, most people recognize strawberries and raspberries, and all of the species in these two groups are edible, though not all are equally palatable. Plants belonging to the deadly nightshade (solanum) family, however, may be more difficult to distinguish from each other and can range in edibility from highly edible to poisonous. Even the most experienced harvester can sometimes make mistakes. It is important to be certain of a species' identification and any special treatment required before eating a wild berry. Serviceberries, for example, are best cooked to neutralize the poisonous cyanide compounds found in their seeds and many types of under-ripe berries can cause digestive upset or even be poisonous. Some rare individuals have an allergic reaction to certain berry species.

Canadian yew (*Taxus canadensis*)

plantings, some of which have naturalized into the wild and have sweet-tasting fruit. It is not recommended that you sample these non-native fruits without a positive identification. Examples of common poisonous berries are the European lily-of-the-valley and any ornamental yew species.

Finally, some people believe that it is OK to eat berries that they see birds and wildlife enjoying. That is simply not the case so don't test this flawed bit of folklore! Likewise, the fact that a plant has edible fruit does not mean that the plant itself is edible.

Pay attention to where you are harvesting. Fruit growing along the edge of a busy highway or near an

It is also important to know which parts of the berry are edible. For example, while the fleshy "berry" (it is really an "aril") of the Canadian yew tree is considered edible, eating this fruit is not recommended as the small hard seed contained inside is so deadly poisonous that ingesting even a few can cause death! As a general rule, most of our native berry species taste good and are edible. Those berries that have a bitter, astringent or unpalatable taste are telling us that they are toxic or poisonous and that we should not be eating them. These species tend to rely on birds, rather than humans, to eat the fruit and distribute the seeds. The exceptions to these guidelines are the many introduced ornamental plants in our gardens and municipal

Saskatoon berry (*Amelanchier alnifolia*)

Pin cherry (*Prunus pensylvanica*)

industrial area could be contaminated with heavy metals or other pollutants. Municipal plantings might look delicious, but they may be sprayed with pesticides and you might not be welcome to harvest the fruit if it has an ornamental value. Please also remember to harvest on public, not private, lands unless you have received permission from the property owner.

Many plants have developed very effective protective mechanisms. Thorns and stinging hairs discourage animals from touching, let alone eating, many plants. Bitter, often irritating and poisonous compounds in leaves and roots repel grazing animals. Many protective devices are dangerous to humans. The "Warning"

boxes throughout the book include notes of potential hazards associated with the plant(s) described. Hazards can range from deadly poisons to spines with irritating compounds in them. These "Warning" boxes may also describe poisonous plants that could be confused with the species being discussed in the account.

The fine line between delicious and dangerous is not always clearly defined. Many of the plants that we eat every day contain toxins and almost any food is toxic if you eat too much of it. Personal sensitivities can also be important. People with allergies may die from eating common foods (e.g., peanuts) that are harmless to most of the population. Most wild plants are

Thimbleberry (*Rubus parviflorus*)

not widely used today, so their effects on a broad spectrum of society remain unknown.

As with many aspects of life, the best approach is "moderation in all things." Sample wisely—when trying something for the first time, take only a small amount to see how you like it and how your body reacts.

No Two Plants Are the Same

Wild plants are highly variable. No two individuals of a species are identical and many characteristics can vary. Some of the more easily observed characteristics include the colour, shape and size of stems, leaves, flowers and fruits. Other less obvious features,

such as sweetness, toughness, juiciness and concentrations of toxins or drugs, also vary from one plant to the next.

Many factors control plant characteristics. Time is one of the most obvious. All plants change as they grow and age. Usually, young leaves are the most tender, and mature fruits are the largest and sweetest. Underground structures also change throughout the year.

Habitat also has a strong influence on plant growth. The leaves of plants from moist, shady sites are often larger, sweeter and more tender than those of plants on dry, sunny hillsides. Berries may be plump and juicy one year, when shrubs have had plenty of moisture, but they can become dry and wizened during a drought. Without the proper nutrients and environmental conditions, plants cannot grow and mature.

Finally, the genetic make-up of a species determines how the plant develops and how it responds to its environment. Wild plant populations tend to be much more variable than domestic crops, partly because of their wide range of habitats, but also because of their greater genetic variability. Humans have been planting and harvesting plants for millennia, repeatedly selecting and breeding plants with the most desirable characteristics. This process has produced many highly productive cultivars— trees with larger, sweeter fruits, potatoes with bigger tubers and sunflowers with larger, oilier seeds.

These crop species are more productive, and they also produce a specific product each time they are planted. Wild plants are much less reliable.

Wild species have developed from a broader range of ancestors growing in many different environments, so their genetic make-up is much more variable than that of domestic cultivars. One population may produce sweet, juicy berries while the berries of another population are small and tart; one plant may have low concentrations of a toxin that is plentiful in its neighbour. This variability makes wild plants much more resilient to change. Although their lack of stability may seem to reduce their value as crop species, it is one of their most valuable features. Domestic crops often have few defences and must be protected from competition and predation. As

Red-osier dogwood (*Cornus sericea*)

fungi, weeds and insects continue to develop immunities to pesticides, we repeatedly return to wild plants for new repellents and, more recently, for pest-resistant genes for our crop plants.

Disclaimer

This book summarizes interesting, publicly available information about many plants in Canada. It is not intended as a "how-to" guide for living off the land. Rather, it is a guide for people wanting to discover the astonishing biodiversity of our useful plants and to connect to our cultural traditions, especially those of the First Nations. Only some of the most widely used species in Canada, and only some of their uses, are described and discussed. Self-medication with herbal medicines is not recommended. Use of plant medicines and consumption of wild foods should only be considered under guidance from an experienced healer/elder/herbalist. As a field guide, the information presented here is limited, and further study of species of interest should be made using other botanical literature. No plant or plant extract should be consumed unless you are absolutely certain of its identity and toxicity and of your personal potential for allergic reactions. The authors and publisher are not responsible for the actions of the reader.

Rough-fruited fairybells (*Prosartes trachycarpa*)

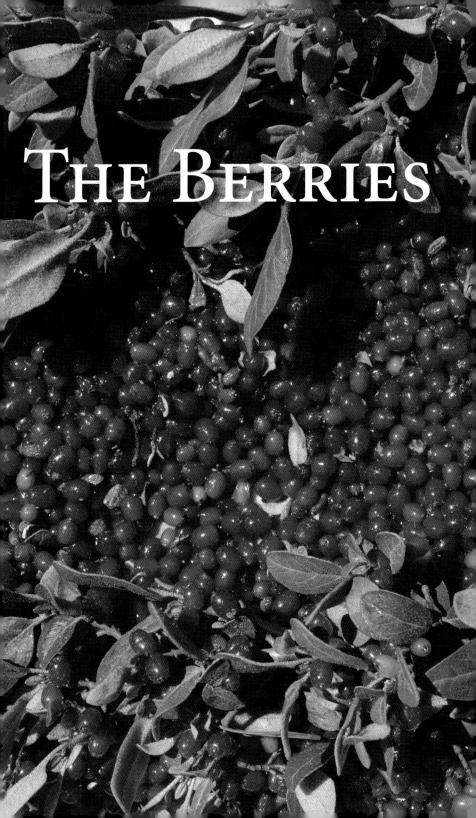

The Berries

Apple *Malus* spp.

Common apple (*M. pumila*)

Several species and cultivars of apple and crabapple are grown for their edible fruits. Though many non-natives have escaped cultivation, wild crabapple is the province's only native species. The fruit of apple trees was highly valued by the Iroquois, Ojibwa, Huron and other Great Lakes tribes. The apples were enjoyed fresh or cooked and were also mashed and dried into small cakes or preserved by the sun or fire drying for winter con-sumption. The tart-tangy dried fruit was rehydrated by soaking in warm water before being cooked into a sauce or mixed with corn bread.

Wild crabapples are very sour tasting, even when fully ripe. They are nice as a limited trail nibble and thirst quencher but are too sour to eat fresh in quantity. The apples have a pleasant aromatic flavour, especially after a frost. When cooked, the fruit is superb in jellies, jams, baking and preserves.

Wild crabapples freeze, dry and jar well. This fruit is very high in pectin, so is useful to mix with other wild fruit such as raspberries and blackberries that are low in pectin and would not set properly without the addition of a commercial pectin product. The tartness of the apples also adds a pleasant "zing" when added to sweeter fruits such as blueberries.

A decoction of wild crabapple roots was used by First Nations to suppress menses. A cold infusion of the bark was prescribed to treat snow blindness and sore eyes. The wood of this species is easy to work and was traditionally used to make tool handles.

EDIBILITY: highly edible

FRUIT: Yellowish-green to red pome with a white flesh, resembling a miniature domesticated apple, round with a depressed shape, fragrant, sometimes waxy, to 6 cm.

SEASON: Flowers May to June. Fruits ripen September to October.

DESCRIPTION: Deciduous shrubs or small trees growing to 12 m tall with reddish-brown, brown or greyish bark. Leaves alternate, simple, ovate to oval, 5–10 cm long, toothed or lobed, turning yellow in fall. Flowers perfect, fragrant, pink to white, in clusters at the ends of side branches. Fruit a green, yellow or red pome.

Common apple (*M. pumila*) is native to central Asia but found across southern Ontario as naturalized relics of cultivation. A small tree 4–12 m tall, thickly branched with brown to greyish bark on the trunk and brownish-

Wild crabapple (*M. coronaria*)

Wild crabapple (*M. coronaria*)

Common apple (*M. pumila*)

red, smooth twigs. Leaves toothed but rarely lobed. Fruit is 2–6 cm in diameter. Often grows in fencerows, old fields, woodlots and even on shores. Also called: wild apple.

Wild crabapple (*M. coronaria*) is a tall shrub or small tree up to 10 m tall with a round-topped crown and spreading limbs. Bark reddish-brown, scaling, longitudinally fissured. Twigs covered in closely matted fine hairs (tomentum) when young, maturing to smooth reddish brown, and developing thorn-like spurs and thorns to 3 cm in their second year. Leaves triangular-ovate to oval, lobed, irregularly toothed. Flowers scented, up to 3 cm across, petals pink in the bud and fading to almost white once open, in flat-clustered groups of 5–6. Fruit yellow or green, up to 25 mm across, on a long stemlet, often persisting on the tree into winter.

Inhabits rich, moist soil in open areas such as streambanks, open woodlands and roadsides in extreme southwestern Ontario. Also called: sweet crabapple, eastern crabapple.

Wild crabapple (*M. coronaria*)

Common apple (*M. pumila*)

Wild crabapple (*M. coronaria*)

Common apple (*M. pumila*)

Junipers *Juniperus* spp.

Common juniper (*J. communis*)

Some indigenous tribes cooked juniper berries into a mush and dried them in cakes for winter use. The berries were also dried whole and ground into a meal that was used to make mush and cakes. In times of famine, small pieces of the bitter bark or a few berries could be chewed to suppress hunger. Dried, roasted juniper berries have been ground and used as a coffee substitute, and teas were occasionally made from the stems, leaves and/or berries, but these concoctions were usually used as medicines rather than beverages. Juniper berries are well known for their use as a flavouring for gin, beer and other alcoholic drinks.

Tricky Marys can be made by soaking juniper berries in tomato juice for a few days and then following your usual recipe for Bloody Marys, but omitting the alcohol. The taste is identical and the drink is non-alcoholic.

Juniper berries can be quite sweet by the end of their second summer on the plant or in the following spring, but they have a rather strong, "pitchy flavour" that some people find distasteful. The berries can be added as flavouring in meat dishes (recommended for venison and other wild game, veal and lamb), soups and stews, either whole, crushed or ground and used like pepper.

Juniper berry tea has been used to aid digestion, stimulate appetite, relieve colic and water retention, treat diarrhea and heart, lung and kidney problems, prevent pregnancy, stop bleeding, reduce swelling and inflammation and calm hyperactivity. The berries were chewed to relieve cold symptoms, settle upset stomachs and increase appetite. Oil-of-juniper (made from the berries) was mixed with fat to make a salve that would protect wounds from irritation by flies. Juniper berries are reported to stimulate urination by irritating the kidneys and will give the urine a violet-like fragrance. They are also said to induce sweating, mucous secretion, production of hydrochloric acid in the stomach and contractions in the uterus and intestines. Some studies have shown juniper berries to lower blood sugar caused by adrenaline hyperglycemia, suggesting that they may be useful in the treatment of insulin-dependent diabetes. The berries also have antiseptic qualities, and studies by the National Cancer Institute have shown that some junipers contain antibiotic compounds that are active against tumours. Strong juniper tea has been used to sterilize needles and bandages, and during the Black Death in 14th-century Europe, doctors held a few berries in the mouth as they believed that this would prevent them from being infected by patients. During cholera epidemics in North America, some people drank and bathed in juniper tea to avoid infection. Juniper tea has been given to women in labour to speed delivery, and after the birth it was used as a cleansing, healing agent.

Juniper berries were sometimes dried on strings, smoked over a greasy fire and polished to make shiny black

Common juniper (*J. communis*)

Common juniper (*J. communis*)

37

beads for necklaces. Some tribes also scattered berries on anthills—the ants would eat out the sweet centre, leaving a convenient hole for stringing the necklaces. Smoke from the berries or branches of junipers has been used in religious ceremonies or to bring good luck (especially for hunters) or protection from disease, evil spirits, witches, thunder, lightning and so on. The berries make a pleasant, aromatic addition to potpourris, and vapours from boiling juniper berries in water were used to purify and deodorize homes affected by sickness or death.

Eastern red cedar (*J. virginiana*)

Common juniper (*J. communis*)

These plants are decorative, particularly in the winter months, and make a hardy and drought-tolerant addition to the ornamental garden. Junipers can be very long-lived, with some specimens recorded as old as 1500 years.

EDIBILITY: edible, but with caution

FRUIT: Small fleshy cones ("berries") are ripe when bluish-purple to bluish-green in colour.

Common juniper (*J. communis*)

Common juniper (*J. communis*)

SEASON: Berries form from May to June on female plants only and mature the following year, but are present on the plant all year round.

DESCRIPTION: Coniferous, evergreen shrubs or small trees with some species creeping low on the ground. Leaves scale-like (1–5 mm) and opposite or needle-like (to 20 mm) and whorled, dark green to yellowish. Male plants produce yellow pollen on cones 5 mm long. Females bear small, 5–9 mm wide berries, first green and maturing to a bluish-purple colour. Grows on open, dry rocky areas and grasslands.

Creeping juniper (*J. horizontalis*)

Common juniper (*J. communis*) grows to 1.5 m tall but is normally lower than this. Growth habit is branching, prostrate, trailing, forming wide mats 1–3 m in size. Leaves (needles) are awl-shaped, dark green above, whitish below, prickly, 12–20 mm long, in whorls of 3. Bark reddish-brown, scaly, thin, shredding. Grows on dry, sandy or rocky open sites and forest edges, gravelly ridges and muskeg from lowland bogs to plains and subalpine zones. Also called: ground juniper.

Creeping juniper (*J. horizontalis*) is a low shrub (seldom over 25 cm tall) with trailing branches. Leaves scale-like, tiny and overlapping in 4 vertical rows lying flat against the branch.

Creeping juniper (*J. horizontalis*)

Eastern red cedar (*J. virginiana*)

Creeping juniper (*J. horizontalis*)

Eastern red cedar (*J. virginiana*)

Grows in sand dunes and dry, rocky soils in woods, pastures and fields.

Eastern red cedar (*J. virginiana*) is a small tree growing to 20 m tall. Leaves green but sometimes turning reddish brown in winter, 1–3 mm long, overlapping by more than a quarter of their length. Grows in upland to low woods, old fields, glades and river swamps in southern Ontario.

Common juniper (*J. communis*)

Hawthorns *Crataegus* spp.

Common hawthorn (*C. monogyna*)

The fruits, or haws, of these plants are edible. The taste of the haws, however, can vary greatly depending on the species, particular tree, time of year and growing conditions. The haws are usually rather seedy, with the flavour described as a range of sweet, mealy, insipid, bitter, astringent or even tasteless. Frosts are known to increase the sweetness of the haws. Historically these berries were eaten fresh from the tree, or dried for winter use. They were also often an addition to pemmican. The cooked, mashed pulp (with the seeds removed) was dried and stored in cakes as a berry-bread, which could be added to soup or eaten with deer fat or marrow. A popular local food of Manitoulin Island, Land of the Haweaters, "hawberries" are commonly prepared and sold as jams, jellies and syrups. Haws are rich in pectin and if boiled with sugar can be a useful aid in getting jams and jellies to set without a commercial pectin product. They can also be steeped to make a pleasing tea or cold drink. The haws of English hawthorn are commonly called "bread and butter berries" in the UK, likely

Black hawthorn (*C. douglasii*)

because of the starchy and somewhat creamy texture of the fruit as you nibble it, especially after a frost.

Hawthorn flowers and fruits are commonly used in herbal medicine as heart tonics, though not all species are equally effective. Studies have supported the use of hawthorn extracts as a treatment for high blood pressure associated with a weak heart, angina pectoris (recurrent pain in the chest and left arm owing to a sudden lack of blood in the heart muscle) and arteriosclerosis (loss of elasticity and thickening of the artery walls). Hawthorn is believed to slow the heart rate and reduce blood pressure by dilating the large arteries supplying blood to the heart and by acting as a mild heart stimulant. However, hawthorn has a gradual, mild action and must be taken for extended periods to produce noticeable results. Hawthorn tea has also been used to treat kidney disease and nervous conditions such as insomnia. Dark-coloured haws are especially high in flavonoids and have been steeped in hot water to make teas for strengthening connective tissues damaged by inflammation. The haws were sometimes eaten in moderate amounts to relieve diarrhea (some indigenous peoples considered them very constipating). The scientific name "*Crataegus*" derives from the Greek *kratos*, which means "strength" and refers to the hard quality and durability of the wood. The common name "hawthorn" derives from the Old English word for a hedge, or "haw"; the species was historically planted and worked into hedgerows where its spiky thorns, branching nature and durable wood made it a formidable and lasting barrier.

EDIBILITY: edible

FRUIT: Fruits are haws, hanging in bunches; small, pulpy, yellow, orange, red, blue or purplish-black pomes (tiny apples) containing 1–5 nutlets.

SEASON: Flowers May to June. Haws ripen late August to September.

DESCRIPTION: Deciduous shrubs or small trees growing 6–11 m tall with strong, straight thorns growing directly from younger branches. Leaves alternate, simple or lobed, generally oval, with a wedge-shaped base and smooth to roughly toothed margins. Flowers whitish to pink, 5-petalled, sometimes unpleasant-smelling, forming showy, flat-topped clusters, from May to June.

Black hawthorn (*C. douglasii*)

Black hawthorn (*C. douglasii*) grows to 11 m tall with 1–2 cm-long thorns at times scattered or absent. Leaves are toothed to shallowly lobed and longer than wide. The haws are about 1 cm long, purplish black in colour. Grows in forest edges, open disturbed sites, lakeshores, streambanks, rocky ridges, thickets and roadsides in lowland to montane zones in southern Ontario west to Lake Superior. Also called: Western hawthorn, thorn apple.

Cockspur hawthorn (*C. crus-galli*) grows 6–10 m tall with a broadly domed or depressed crown and mildly scaly grey-brown bark. Branches horizontal, stiff, wide-spreading. Thorns 2–6 cm long, numerous, straight or slightly curved. Buds dark brown, rounded, often 2–3 together, with one producing a thorn. Leaves mostly unlobed and leathery, widest above the middle with wedge-shaped bases and broad tips. The upper surfaces are dark green and glossy while the undersides are lighter in colour and dull. Flowers hairless, loose, in flat-topped clusters. Haws 8–10 mm wide, green to dull red, egg-shaped to round, often dark-dotted and persisting through winter. Inhabits open, often disturbed areas

on dry rocky ground, open woodlands and pastures in extreme southern Ontario. The Latin name for this species translates as "cock's shin or leg," in reference to the shape and sharpness of the thorns. Also called: cockspur thorn.

Dotted hawthorn (*C. punctata*) grows 6–10 m tall with a broad open crown, often forming thickets. Slender thorns (sometimes slightly curved or branched) 3–6 cm long on trunks and small branches. Leaves longer (2–8 cm) than wide (1–5 cm), often unlobed, widest and sharply single or double-toothed above midleaf, dull green with impressed veins above,

Cockspur hawthorn (*C. crus-galli*)

Cockspur hawthorn (*C. crus-galli*)

Downy hawthorn (*C. mollis*)

Dotted hawthorn (*C. punctata*)

Dotted hawthorn (*C. punctata*)

slightly hairy below. Flowers May to June, unpleasant-smelling. Fruits pear-shaped to spherical, 1–1.5 cm wide, dull red or orange-red, pale-dotted, often persisting through winter. Inhabits open rocky ground in floodplains, clearings and thickets in southern Ontario. Also called: whitehaw.

Downy hawthorn (*C. mollis*) grows up to 10 m in height, with a broad crown and often forming thickets. Thorns few, 2–6 cm long, slender, straight, reddish-brown. Leaves 4–8 cm long, widest at or below midleaf, sharply double-toothed or with 9–11 sharp-

toothed lobes. Flowers May to June in showy, hairy-branched, flat-topped clusters on dwarf shoots. Fruits bright red haws, hairy near the ends, round to pear-shaped, 1–1.5 cm, often with 5 seeds. Inhabits open areas, disturbed sites such as fencelines, pasture edges and roadsides as well as thickets and woody hillsides in southern and eastern Ontario. Also called: red hawthorn.

Common hawthorn (*C. monogyna*) usually grows 5–10 m tall but can reach 14 m. Thorns 1–2 cm long borne on younger stems, bark of older stems is dull brown, sometimes with orange-shaded cracks. Leaves are dark on top, paler underneath, 2–5 cm long, deeply lobed sometimes to the midrib, obovate. Fruits are dark red, often less than 1 cm thick, containing a single, hard nutlet. Introduced to a limited range in southern Ontario but now extending its range from Georgian Bay to the St. Lawrence River; the species is native to Europe, NW Africa and Western Asia. Also called: English hawthorn, May tree.

Fanleaf hawthorn (*C. flabellata*)

Fanleaf hawthorn (*C. flabellata*)

Fleshy hawthorn (*C. succulenta*)

Common hawthorn (*C. monogyna*)

Fanleaf hawthorn (*C. flabellata*) grows 5–6 m tall with a broad crown, often in thickets. Thorns numerous, straight or slightly curved, 5–6 cm long (sometimes 3–10 cm). Leaves often bent backwards, blades hairless when mature, edged with 7–13 sharply toothed lobes. Haws crimson, juicy, thick-fleshed, often persisting through winter. Inhabits open, rocky sites and clearings, pastures and edges of watercourses in southern Ontario up to the Bruce Peninsula and the Ottawa Valley. This species also resembles, and may include, bigfruit hawthorn (*C. macrosperma*). Also called: New England hawthorn.

Fireberry hawthorn (*C. chrysocarpa*) is a shrub or small tree, to 6 m tall, with a crooked, highly branched trunk. Stout branches usually have numerous shiny black thorns, 2–8 cm long. Leaves are dull yellowish green, toothed, with small, triangular lobes. Flowers small, 1–1.5 cm wide, white, blooming May to June. Haws usually green until late summer then turn

Fleshy hawthorn (*C. succulenta*)

Fireberry hawthorn (*C. chrysocarpa*)

margins with acute, shallow lobes. Haws bright red, glossy, 7 mm–1.2 cm long. Grows in thickets, pastures and woodland edges in southern Ontario and less commonly around Lake Superior. The Latin name for this species means "with juicy flesh." Also called: long-spined hawthorn, succulent hawthorn.

Scarlet hawthorn (*C. pedicellata*) grows to 10 m, with a conical and compact crown. Thorns smooth, stout, shiny, 2–6 cm long, usually slightly curved. Leaves 6–9 cm long, widest below midleaf, ovate to almost round, rough-hairy when young, sharply double-toothed, 7–11 shallow lobes. Haws bright red, thick-fleshed, spherical, 1–1.4 cm across. Inhabits disturbed sites and open areas such as pasture and field edges in extreme southern Ontario. Also called: syn. *C. aulica*, *C. allwangeriana*.

deep red, often persisting through winter. Found on open, gravelly sites near water in southern and central Ontario. Also called: golden-fruited hawthorn.

Fleshy hawthorn (*C. succulenta*) is a shrub or multi-stemmed, shrubby tree, growing 6–8 m tall with a broadly domed crown. Thorns blackish, glossy, strong, up to 8 cm long. Leaves elliptic to ovate, 3–7 cm long, with toothed

Scarlet hawthorn (*C. pedicellata*)

47

Mountain Ashes *Sorbus* spp.

European mountain ash (*S. aucuparia*)

The bitter-tasting fruits of these trees are high in vitamin C and can be eaten raw, cooked or dried. These fruit were consumed by a number of First Nations groups but many considered them inedible. After picking, these berries were sometimes stored fresh underground for later use. They were also added to other more popular berries or used to marinate meat such as marmot. The green berries are too bitter to eat, but the ripe fruit, mellowed by repeated freezing, is said to be tasty enough.

These species have been used to make jams, jellies, pies, ale and also bitter-sweet wine, and the fruit is also enjoyed cooked and sweetened. In northern Europe, the berries, which can be quite mealy, were historically dried and ground into flour, which was fermented and used to make a strong liquor. A tea made of the berries is astringent and has been used as a gargle for relieving sore throats and tonsillitis.

European mountain-ash fruit has been used medicinally to make teas for treating indigestion, hemorrhoids, diarrhea and problems with the urinary tract, gallbladder and heart. Some indigenous peoples rubbed the berries into their scalps to kill lice and treat dandruff. European mountain-ash is a popular ornamental tree, and the native mountain ashes make attractive garden shrubs, easily propagated from seed sown in autumn. The scarlet fruit can persist throughout winter, and the bright clusters of fruit attract many birds.

EDIBILITY: edible, but not great

FRUIT: Fruits berry-like pomes, about 1 cm long, hanging in clusters.

European mountain ash (*S. aucuparia*)

SEASON: Blooms June to July. Fruits ripen August to September.

DESCRIPTION: Clumped, deciduous shrubs or slender trees with smooth, brownish bark. Numerous lenticels (raised ridges that are actually breathing pores) occur on young bark, turning grey and rough with age. Leaves alternate and pinnately compound with 9–17 sharply-toothed, lance-like leaflets 5–10 cm in length, darker above, paler below. Flowers white, about 1 cm across, 5-petalled, forming flat-topped clusters, 9–15 cm wide, smelly. Grows in sun-dappled woods, rocky ridges and forest edges, preferring moist areas and full sun.

American mountain ash (*S. americana*)

American mountain ash (*S. americana*) is a small tree, to 10 m tall, much less hairy than European mountain ash. The buds, leaf stems and leaves hairless or nearly so, with bright coral-red fruits 4–6 mm in diameter. The leaflets are generally longer and narrower than those of showy mountain ash, which also has larger fruits and blooms slightly later in the season. European mountain ash has a hairy lower leaf surface and buds and larger fruits and flowers. American mountain ash grows in moist sites along swamps, wet forests and rocky hillsides in south and eastern Ontario extending north to James Bay.

European mountain ash (*S. aucuparia*) is a commonly planted ornamental tree, to 15 m tall, with white-hairy buds, leaf stems and leaves (underneath at least) and orange to red fruit of about 1 cm in diameter in large clusters up to 20 cm across. This Eurasian species is widely cultivated and just as widely escaped. Its leaflets are generally shorter (3–5 cm) than those of native species. Also called: Rowan tree.

Showy mountain ash (*S. decora*)

Showy mountain ash (*S. decora*) grows to 10 m tall with a short, rounded crown. Branches are reddish-brown, think-skinned with dark gummy buds. Compound leaves 10–25 cm long with 13–17 bluish-green leaflets each 3–8 cm long. Inhabits rocky shores of watercourses and lakes in all except extreme northern Ontario. Very similar to American mountain ash (see the latter species description for differences). Also called: northern mountain ash, dog-berry.

Wild Roses *Rosa* spp.

Prickly wild rose (*R. acicularis*)

Most parts of rose shrubs are edible and the fruit (hips), which remain on the branches throughout winter, are available when most other species have finished for the season. The hips can be eaten fresh or dried and are most commonly used in tea, jam, jelly, syrup and wine. Usually only the fleshy outer layer is eaten (see Warning). Because rose hips are so seedy, some indigenous peoples considered them as famine food rather than regular fare. Rose petals have a delicate rose flavour with a hint of sweetness and may be eaten alone as a trail nibble, added to teas, jellies and wines or candied. Adding a few rose petals to a regular salad instantly turns it into a delicious gourmet conversation piece, and guests are often surprised at how delicate and sweetly delicious the petals taste. Do not eat commercial rose petals, however, as they are often sprayed with chemicals.

Rose hips are rich in vitamins A, B, E and K and are one of our best native sources of vitamin C—three hips can contain as much as a whole orange! During World War II, when oranges could not be imported, British and Scandinavian people collected hundreds of tonnes of rose hips to make a nutritional syrup. The vitamin C content of fresh hips varies greatly, but that of commercial "natural" rose hip products can fluctuate even more.

Rose petals have been taken to relieve colic, heartburn and headaches. They were also ground and mixed with grease to make a salve for mouth sores or mixed with wine to make a medicine for relieving earaches, toothaches and uterine cramps. Dried rose petals have a lovely fragrance and are a common ingredient in potpourri. Rose sprigs were hung on cradle boards to keep ghosts away from babies, and on

Multiflora rose (*R. multiflora*)

the walls of haunted houses and in graves to prevent the dead from howling. Some native roses can hybridise with each other, resulting in offspring that have mixed traits.

EDIBILITY: edible

FRUIT: Fruits scarlet to purplish, round to pear-shaped, berry-like hips, 1.5–3 cm long, with a fleshy outer layer enclosing many stiff-hairy achenes.

SEASON: Blooms June to August. Hips ripen August to October.

DESCRIPTION: Thorny to prickly and often bristly, deciduous or persistent shrubs and woody vines 30 cm–2 m tall. Spines generally straight (introduced rose species tend to have curved spines). Leaves alternate, pinnately divided into 3–11 oblong, toothed leaflets, generally odd in number. Flowers of native species light pink to deep rose, 5-petalled, fragrant, usually growing at the tips of branches. Grows in a wide range of habitat from dry rocky slopes, forest edges, woodlands and clearings to roadsides and streamsides at mid- to low-level elevations.

Multiflora rose (*R. multiflora*) is a sprawling, prickly shrub up to 5 m tall that can form dense impenetrable thickets. Leaves up to 8–11 cm long consisting of 5–11 sharply toothed leaflets. Flowers in May or June, clustered at end of branches with small bright to deep red hips less than 8 mm across. The hips are less tasty than other species but are highly prized by different types of birds. Introduced from Asia as a rootstock for ornamentals, this plant now grows in old fields, open woodland and forest edges in southern Ontario. Also called: baby rose.

Prickly wild rose (*R. acicularis ssp. sayi*) grows to 1.5 m tall but is often shorter, with bristly, prickly branches, 5 or 7 leaflets up to 5 cm long and small clusters of 5–7 cm-wide flowers or 1–2 cm-long bright red hips. Grows in open woods, thickets and on rocky slopes as well as clay, sand and riverbanks throughout the province.

Rugosa rose (*R. rugosa*) is an erect shrub, 1–1.5 m tall and wide, with fine, dense thorns up to 1 cm long. Deciduous compound leafs of 5–9 dark green leaflets with a distinctively rough ("rugose") surface. Sweet-scented, white to purple flowers borne singly, 5–9 cm across and tomato-like red

Prickly wild rose (*R. acicularis*)

hips that are wider than long. An introduced rose from Asia, it grows well in sandy soils (dunes, beaches and roadsides) but not in poorly drained areas of southern Ontario. Also called: beach tomato.

Smooth wild rose (*R. blanda*) grows to 1.5 m tall. Branches and stems are unarmed with the exception of a few slender, straight prickles near the base of new shoots. Branchlets reddish-purple. Leaves with 5–7 leaflets (sometimes 9), oval to egg-shaped, 1–4.5 cm long, sharply toothed margins below leaf middle. Pair of stipules (leaf-like bracts) at base of stalk. Flowers pink, saucer-shaped, 2–3 cm long, most often single but sometimes more, clustered. Hips red, smooth, round to pear- or egg-shaped, 1–1.5 cm in diameter. Inhabits open forests, fields, hedgerows, shorelines and clearings as well as moist to dry, sandy to loamy upland sites in all but the most northern regions of Ontario.

Swamp rose (*R. palustris*) is a many branched, arching shrub to 2.5 m tall. Stems numerous, bushy-branched and

Prickly wild rose (*R. acicularis*)

red-brown. Prickles stout, hooked and paired. Leaves of 7 (but occasionally 5 or 9) narrow leaflets up to 6 cm long. Flowers pink and delightfully fragrant from June to August. Hips very prickly when green, less so when ripe. Inhabits open, moist to wet sites in southern Ontario, from the southeastern tip of Lake Superior to Québec.

> **WARNING:** *The dry inner "seeds" (achenes) of the hips are not palatable and their fibreglass-like hairs can irritate the digestive tract and cause "itchy bum" if ingested. As kids, we used to make a great old-fashioned itching powder by slicing a ripe hip in half and scraping out the seeds with these attached hairs. Spread this material to dry, then swirl it in a bowl, and the seeds will drop to the bottom. Skim off the fine, dry hairs—this is your itching powder, guaranteed to work. While all members of the Rose family have cyanide-like compounds in their seeds, these are destroyed by drying or cooking.*

Prickly wild rose (*R. acicularis*)

Smooth wild rose (*R. blanda*)

Swamp rose (*R. palustris*)

Rosehip Jelly

Makes 8 x 1 cup jars

2 lbs whole rosehips · 2 lbs apples · 5 cups water · juice of 1 lemon
6 to 8 cloves · small cinnamon stick · white sugar

Carefully wash rosehips and apples. *While any ripe rosehips will work, in my experience those of the swamp rose have the most superior flavour. Slightly unripe apples work best for this recipe as they have a higher pectin content than ripe fruit does.* Core apples and chop roughly. Place the fruit in separate cooking pans with 2½ cups of water in each pan. Add lemon juice, cloves and cinnamon to the pan containing the rosehips. Bring both pans gently to the boil, then reduce heat and simmer until the fruit is soft and pulpy. Place the contents of both pans together in a jellybag and allow the juice to strain through overnight into a clean bowl. *If you want a perfectly clear jelly, do not press or squeeze the bag.*

In the morning, measure the strained liquid and allow for 2 cups of sugar to every 2½ cups of juice. Place the juice and sugar in a thick-bottomed cooking pan. *A thick-bottomed pan is important, because a thin-bottomed pan will get too hot and scald the jelly.* Bring to the boil, stirring and being careful to scrape the bottom of the pan, until the sugar is dissolved. Boil until setting point is reached (when you take some of the liquid on a wide-lipped spoon, blow on it to cool, then start to pour it off the side of the spoon and it gels together). Meanwhile, prepare 8 x 236 mL jars and lids (wash and sterilize jars and lids, and fill jars with boiling water; drain just before use).

Pour the hot jelly into clean, hot, sterilized jars. Seal the jars and place out of the sun to cool.

Black Chokeberry   *Photinia melanocarpa*

Also called: black chokecherry (*Aronia melanocarpa*)

Black chokeberry (*P. melanocarpa*)

The fruit of this species reportedly has a good flavour but can be astringent and bitter, especially when not fully ripe. It does, however, sweeten after a frost and is also much improved by cooking and adding sugar to taste. These dark-coloured fruits can grow profusely on the bush and are easy to gather in quantity. They make a tasty and good-looking jelly, either alone or mixed with other fruits. These pomes are rich in pectin and can be added to low-pectin fruits such as blueberries when making jam or jelly to get a stronger set. The seeds are so small and soft as to be practically unnoticeable so do not need to be strained out. An infusion of the ripe fruit was traditionally used as a cold remedy, and the berries were also mixed with other fruit when making pemmican.

EDIBILITY: edible

FRUIT: Fleshy round, berry-like pome in small clusters, 6–10 mm in diameter, purple to black in colour, bottom end of fruit has a distinctive star pattern. Sometimes persist on the plant over winter.

SEASON: Flowers May to June. Fruits ripen July to September.

DESCRIPTION: Perennial deciduous shrub to 2.5 m tall and 3 m wide with a leggy growth habit. Produces suckers and can form thickets. Stems nearest the base are leafless. Branchlets purple to grey-brown, sometimes hairy. Young stems reddish brown, older stems have prominent lenticels (horizontal breathing pores in the bark). Leaves glossy, bright green, 2–9 cm long and 1–4 cm wide, alternate, oval to elliptic, simple, tapering at base, sharp-pointed or tapered at tip. Leaf stalks 2–10 mm long. Upper surface hairless, dark green with dark hair-like

Black chokeberry (*P. melanocarpa*)

glands in a row along the midrib (viewed with a magnifying glass), paler and sometimes hairy below, turning red in fall. Fine gland-tipped teeth on margins. Flowers 4–6 mm long, 5-petalled, white, in clusters of 5–15 at branch ends. Inhabits wet or moist forest edges, thickets, swamps, wet to dry clearings, streambanks and lakeshores. Common throughout southern Ontario and rare north and west of Lake Superior, this shrub prefers a moist habitat but will tolerate a wide range of conditions from moist to dry.

Black chokeberry (*P. melanocarpa*)

Cherries *Prunus* spp.

Pin cherry (*P. pensylvanica*)

Cherries may be eaten raw as a tart nibble, but the cooked or dried fruit is much sweeter and additional sugar further improves the flavour. The fruit can be cooked in pies, muffins, pancakes and other baking, or strained and made into jelly, syrup, juice, sauce or wine. It seldom contains enough natural pectin to make a firm jelly, however, so pectin must be added (see hawthorns for a natural source rather than store-bought preparations). Although wild cherries are small compared to domestic varieties, they can be collected in large quantities. However, pitting such small fruits is a tedious job, especially since they are too tiny to use with a modern cherry-pitting tool.

When in flower, the pin cherry tree is a dramatic and sweet-scented pleasure to behold so is well worth considering for the ornamental garden. Tradition-

Pin cherry (*P. pensylvanica*)

ally, wild cherry fruit were eaten fresh or cooked, or were dried and then powdered to store for winter use.

Wild cherry fruit are worth considering in landscaping as these are a favourite food for many mammals such as chipmunks, rabbits, mice, deer, elk and moose, and birds such as robins and grouse. The wood of wild black cherry is sought after by furniture makers and fine craftspeople for its rich, red colour and durability. It has traditionally been used to make musical instruments, furniture, frames, paneling and tool handles. Black cherry was one of the first North American trees to be introduced to England in 1629.

EDIBILITY: highly edible

FRUIT: Fleshy drupes (cherries) with large stones (pits), ranging in colour from red to blackish purple to black.

SEASON: Flowers April to June. Fruits ripen July to August.

DESCRIPTION: Deciduous shrubs or small trees growing 1–25 m tall. Trunk and branches reddish-brown to nearly black, often shiny, with raised horizontal "breathing" pores (lenticels) prominent in stripes on the trunk and larger branches. Leaves alternate, smooth, fine-toothed, sharp-tipped and 3–10 cm long. Flowers white or pinkish, about 1 cm across, 5-petalled, forming small, flat-topped clusters.

Pin cherry (*P. pensylvanica*) grows to 12 m tall. Leaves slender with long-tapering points, lance-shaped, sharp-toothed to 15 cm in length. White flowers in clusters, 5–7 along twigs. Fruit bright red cherries, 4–8 mm long, thin sour flesh. Grows through-

Pin cherry (*P. pensylvanica*)

out Ontario in moist thickets, woods, riverbanks, forest edges, disturbed areas and clearings from sea level to subalpine zones. Also called: bird cherry, fire cherry, Pennsylvania cherry.

Sand cherry (*P. pumila*) is a low, slender shrub to 1 m in height that often spreads close to the ground. Leaves to 10 cm long and 3 cm wide, dark green above, paler below and often leathery with only finely toothed edges above the middle. Flowers white in small groups of 2–4, produce purple or blackish cherries 8–15 mm in diameter with an acidic, astringent taste. Found across the province, particularly in the south but also northward to James Bay, in well-drained sandy and rocky habitats. Also called: dwarf cherry, low sand cherry.

WARNING: *Cherry leaves, bark, wood and seeds (stones) contain hydrocyanic acid and therefore can cause cyanide poisoning. The flesh of the cherry is the only edible part. The stone should always be discarded, but cooking or drying destroys the toxins. Cherry leaves and twigs can be poisonous to browsing animals.*

Wild black cherry (*P. serotina*) is a tall shrub or tree growing 20–30 m tall with a rounded crown. Trunk straight with reddish- or blackish-brown bark covered in prominent lenticels. Young bark smooth, shiny. Mature bark rough with squared scales and often fissured on larger trees. Branches arching with drooping tips, strongly smelling when broken. Leaves 5–15 cm long and less than half as wide, shiny dark green above, paler below with fine white or rust-coloured hairs. Flowers white, on 5 mm stalks, in clusters of 10–15, hanging from the tips of new shoots.

Pin cherry (*P. pensylvanica*)

Wild black cherry (*P. serotina*)

Sand cherry (*P. pumila*)

Wild black cherry (*P. serotina*)

Sand cherry (*P. pumila*)

Fruits reddish to black cherries, purplish flesh with a single stone, hanging in elongated clusters. Inhabits open woodlands, rocky areas and disturbed sites such as riverbanks and logged areas. Also called: timber cherry, black chokecherry, rum cherry, wine cherry.

Wild black cherry (*P. serotina*)

Chokecherry *Prunus virginiana*

Also called: wild cherry

Chokecherry (*P. virginiana*)

Chokecherries were among the most important and widely used berries by First Nations across Canada, and dried chokecherries were a valuable trade item for some tribes. The fruits were highly regarded and were collected after a frost (which makes them much sweeter) and were dried or cooked, often as an addition to pemmican or stews. Large quantities were gathered, pulverized with rocks, formed into patties about 15 cm in diameter and 2 cm thick and dried for winter use. Chokecherries were most commonly dried with the pits intact (a process that destroys the toxic hydrocyanic acid in the pits) and could also be stored when picked as branches for several months if kept in a cool, dry place.

Chokecherry (*P. virginiana*)

Today, chokecherries are used to make beautifully coloured jelly, syrup, sauce and beer as well as wine. The raw cherries are sour and astringent, particularly if they are not fully ripe, so they cause a puckering or choking sensation when eaten—hence the common name "choke cherry." One unimpressed early European traveller in 1634 is reported to have written that "chokecherries so furre the mouthe that the tongue will cleave the roofe, and the throate wax hoarse"! After the cherries have been cooked or dried, however, they are much sweeter and lose their astringency. Dried, powdered choke cherry flesh was taken to improve appetite and relieve diarrhea and bloody discharge in the bowels. The bark and roots were prepared and also used as an appetite stimulant as well as for treating coughs, tuberculosis and as a sedative and vermifuge.

EDIBILITY: highly edible (when fully ripe, after a frost or sweetened)

Chokecherry (*P. virginiana*)

Chokecherry (*P. virginiana*)

Chokecherry (*P. virginiana*)

FRUIT: Red, black to mahogany-coloured, shiny, growing in heavy and generous clusters 6–12 mm wide. Some reports indicate that the red fruit have a nicer flavour than the darker-coloured ones.

SEASON: Flowers May to June. Fruits ripen August to September.

DESCRIPTION: Deciduous shrubs or most often small trees growing to 8 m tall. Bark smooth, greyish marked with small lenticels. Leaves alternate, 3–10 cm long, broadly oval, finely sharp-toothed, with 2–3 prominent glands near the stalk tip. Flowers creamy white, 10–12 mm across, 5-petalled, forming bottlebrush-like clusters 5–15 cm long. Chokecherry is a fast-growing tree and short-lived

Chokecherry (*P. virginiana*)

species that inhabits sunny disturbed sites in deciduous woods, open sites, streams and forest edges across the province, especially in southern regions.

WARNING: *Like other species of* Prunus *and* Pyrus, *all parts of the chokecherry (except the flesh of the fruit) contain cyanide-producing glycocides. There are reports of children dying after eating large amounts of fresh chokecherries without removing the stones. Cooking or drying the seeds, however, appears to destroy most of the glycocides. Chokecherry leaves and twigs are poisonous to animals.*

63

Sumacs *Rhus* spp.

Staghorn sumac (*R. typhina*)

The showy, red fruit clusters of sumacs are beautiful to look at and can be made into a refreshing and pretty, pink or rose-coloured drink with a lemon-like flavour. Crush the berries then soak the mash in cold water before straining to remove the fine hairs from the fruit and other debris. The juice is best sweetened with sugar and served cold. Don't pour on hot water or boil the fruit, however, as the heat releases tannins and produces a bitter-tasting liquid. The fruits have also been used to make jellies or lemon pies. The tangy lemon flavour of sumac fruit (which really comes from the hairs covering the seeds) is a common ingredient in Middle Eastern cuisine. When chewed as a trail nibble, sumac fruits relieve thirst and leave a pleasant taste in the mouth.

Beyond food, some First Nations people boiled the fruits to make a wash to stop bleeding after childbirth. The berries, steeped in hot water, made a medicinal tea for treating diabetes, bowel problems and fevers. This tea was also used as a wash for ringworm, ulcers and skin diseases such as eczema.

Sumac is a very decorative and hardy species that provides an interesting fall and winter garden display. It does tend to sucker, though, so can get invasive in the garden if not kept in check.

EDIBILITY: edible

FRUIT: Fruits reddish, densely hairy, berry-like drupes, 4–5 mm long, in persistent, fuzzy clusters.

SEASON: Flowers May to July. Fruits ripen July to August, often remaining through the winter.

DESCRIPTION: Deciduous shrubs or small trees, 1–6 m tall, usually forming thickets. Branches exude milky juice when broken. Leaves trifoliate (consists of three leaflets) or pinnately divided into 11–31 lance-shaped, 5–12 cm long, toothed leaflets, bright red in autumn. Flowers cream-coloured to greenish yellow, about 3 mm across, with 5 fuzzy petals, forming dense, pyramid-shaped, 10–25 cm-long clusters.

Fragrant sumac (*R. aromatica*) is a low shrub up to 1.5 m high that may form mounds or thickets of aromatic foliage. Leaves trifoliate with the terminal one 5–7 cm long and smaller lateral ones, coarsely toothed. Small yellowish flowers in dense clusters in spring followed by fuzzy, reddish fruits in late summer. Restricted to the southern edge of the Canadian Shield, from Manitoulin Island to the Ottawa Valley, this species prefers dry, sandy or rocky environments.

Smooth sumac (*R. glabra*) grows 1–3 m tall with hairless twigs and leaves; buds with whitish hairs. Leaves up to 30 cm long, compound with 5–14 pairs of lanceolate leaflets plus a terminal one on a stalk. Flowers yellow-green in large terminal clusters. Grows on dry forest openings, prairies, fencerows, roadsides and burned areas in southwestern and southeastern Ontario.

Staghorn sumac (*R. typhina*) grows to 6 m tall with flat-topped crowns with a similar leaf structure as smooth sumac but can reach 50 cm in length. Branches have velvety hair; leaflets have hairy, reddish central stalks; buds densely hairy. Colonizes in open, sandy or rocky areas in southeastern Ontario.

Smooth sumac (*R. glabra*)

Staghorn sumac (*R. typhina*)

Indian Lemonade

Makes 8½ cups

Enhance this beautiful pink "lemonade" by adding ice cubes (or frozen blueberries) and green mint sprigs.

3 cups dried and crumbled sumac flower spikes
8 cups water • sugar to taste

Pick through the dried flower spikes to remove any dirt or debris. Crumble the red "berries" off the main spike and place them in a jug. Pour the cold water over the berries, mash the mixture with a wooden spoon or potato masher, then let sit for at least an hour. *Do not heat this mixture because it will alter the taste of the sumac.* Strain the liquid through a cheesecloth, jellybag or fine-mesh sieve, and add sugar to taste.

Smooth sumac (*R. glabra*)

Blackberries *Rubus* spp.

Common blackberry (*R. allegheniensis*)

Blackberries were traditionally gathered by indigenous peoples and are still widely enjoyed today. They were typically picked in large quantities and eaten fresh or stored (usually dried, either alone or with other fruit) for winter. First Nations used the green twigs of common blackberry to make a black dye and the fruit to make a purple dye. A traditional method of eating the berries was to combine them with other berries or with oil (sometimes whipped) and meat. Valued and marketed for their high antioxidant content, blackberries today are enjoyed in many different ways: on their own, with cream or yoghurt and sugar, in pies and sauces, as jam or jelly or as drinks such as cordial, juice or wine. Blackberries are widely cultivated across Canada for their delicious fruits, and modern thornless cultivars are readily available for the home gardener.

Blackberries hybridise readily so identification can sometimes be difficult. Blackberry-raspberry crosses, such as loganberries and boysenberries, are extremely flavourful and should be more popular and widely known than they are. The word "bramble" comes from the Old English "braembel" or "bro," which means "thorny shrub."

EDIBILITY: highly edible

FRUIT: Juicy red to black drupelets aggregated into clusters.

SEASON: Flowers May to July (sometimes into August in moist, shady or cool spots). Fruits ripen June to September.

DESCRIPTION: Prickly, perennial shrubs, often arching or trailing, branches 0.5–5 m long. Leaves alternate, compound with 5 leaflets arranged palmately (like the fingers on a hand) on first-year branches and 3 leaflets or single blades on flowering branches. Flowers white to pinkish, 5-petalled. Fruit fall from the shrub with the fleshy receptacle intact (i.e., the blackberries have a solid core).

Common blackberry (*R. allegheniensis*)

Common blackberry (*R. allegheniensis*)

Smooth blackberry (*R. canadensis*)

Common blackberry (*R. alleghenien-sis*) is a medium to large erect shrub, to 3 m tall. Stems have stiff, straight bristles, and both stems and leaves are glandular-hairy, usually with 5 palmate leaflets. Fruit up to 25 mm long. Inhabits moist to dry open areas in forests, clearings and roadsides of southeastern Ontario. Also called: Allegheny blackberry, high-bush blackberry.

Smooth blackberry (*R. canadensis*) grows to 2 m. Older canes ridged, brown or reddish-purple with or without scattered prickles. Leaves similar to common blackberry but glabrous (without hair). Fruit round to thimble-shaped drupelets, 12 mm long, hard to separate from their core. Inhabits open rocky areas, streambanks, lakeshores and roadsides. Also called: Canada blackberry.

COMPARE: *Blackberries and their relatives (raspberry, salmonberry, thimbleberry and cloudberry) are all closely related members of the genus* Rubus. *The best way to distinguish blackberries from raspberries is by looking at their fruits: if they are hollow like a thimble, they are raspberries, and if they have a solid core, they are blackberries. Before and after fruiting, the plants can be differentiated according to leaf type on non-flowering branches; palmate with 5 leaflets for blackberries and trifoliate for raspberries.*

Smooth blackberry (*R. canadensis*)

Blackberry Syrup

Makes 3 x 1 cup jars

This fruity syrup makes a delicious warm or cold drink and is recommended for relieving the symptoms of the common cold. Add 1 Tbsp to 1 cup of hot water.

1 lb blackberries
1 cup white wine vinegar
1 cup sugar
4 Tbsp honey

Place the clean fruit into a glass jar and pour the vinegar over top. Leave to stand for at least 24 hours, stirring and crushing the fruit regularly to extract the juices. Strain the liquid into a large saucepan and bring to the boil. Add the sugar, stirring until it's all dissolved. Add honey, stir well. Bring back to the boil, and boil hard for 5 minutes. Allow to cool completely. Pour into sterilized jars while still hot, and seal. Alternatively, you can pour the cooled liquid into ice cube trays in the freezer.

TIP

If you are pouring very hot liquid into a sterilized glass jar that has cooled, the sudden heat can cause the jar to crack. Avoid this problem by first pouring in a few tablespoons of the hot liquid and waiting 10 seconds for the heat to spread, then filling the rest of the jar.

Berry Blackberry Cordial

Makes approximately 4 x 1 cup jars if 8 cups of fruit are used

up to 8 cups freshly picked blackberries (or other juicy berries such as raspberries or thimbleberries, or a combination of berries) • white vinegar • sugar

Carefully pick through the fruit to remove any debris or insects. Be particularly wary of stink bugs, which are about 1 cm in size, green to brownish in colour, flat-backed with a hard carapace, and emit a rank stench if bitten into: they will ruin the entire batch of cordial!

Place the berries in a large glass jar and crush somewhat firmly with a potato masher. Pour enough white vinegar into the jar to just barely cover the fruit mash (roughly an 8:1 ratio). Stir vigorously, put a firm lid on the jar, then let it sit somewhere warm out of direct sunlight for 1 week, stirring once a day.

After a week, strain the mixture overnight through a jellybag. *Resist squeezing it or you will push solids through the bag, resulting in a cloudy end-product with sediment.* The leftover fruit mash can be used in muffins or pancakes.

Measure out the resulting juice into a thick-bottomed saucepan and add 1 cup white granulated sugar for every 1 cup of juice. Slowly bring to the boil to fully dissolve the sugar. Let cool and place in washed, sterilized Mason-type jars for storage. Other glass containers such as maple syrup bottles with rubber-sealed tops also work well.

To make the cordial, mix the concentrate in a 6:1 ratio with cold water. Garnish with a sprig of fresh mint or some frozen whole berries.

Raspberries *Rubus* spp.

Arctic dwarf raspberry (*R. arcticus*)

Wild raspberries are some of our most delicious native berries and are fabulous eaten fresh from the branch or made into pies, cakes, puddings, cobblers, jams, jellies, juices and wines. Because the cupped fruit clusters drop from the receptacle when ripe, these fruits are soft and easily crush to a juicy mess when gathered.

Raspberries were a popular and valuable food of indigenous peoples and were often gathered and processed into dried cakes either alone or with other berries for winter use. These cakes were reconstituted by boiling, or eaten as an accompaniment to dried meat or fish. Fresh or dried leaves of this species have been used to make tea, and the flowers make a pretty addition to salads.

Raspberry leaf tea and raspberry juice boiled with sugar were gargled to treat mouth and throat inflammations. Coureurs de bois used the soft, maple leaf–like leaves of flowering raspberry to protect and cushion their feet in leather shoes. Containing high levels of polyphenolic antioxidants, raspberries are often touted as "superfoods."

EDIBILITY: highly edible

FRUIT: Fruits juicy, red to black drupelets aggregated into clusters that fall from the shrub without the fleshy receptacle.

SEASON: Flowers June to July. Berries ripen July and August.

DESCRIPTION: Armed or unarmed, perennial shrubs or herbs, 15 cm–4 m tall. Leaves deciduous, lobed or compound (divided into leaflets). Flowers white to pink.

Arctic dwarf raspberry (*R. arcticus*) is a small plant growing only 5–15 cm tall. Stems erect, lack prickles, finely hairy, slender. Flowering branches with 2–3 leaves, solitary (rarely 2) terminal dark pink flower, erect, sometimes woody at the base. Leaf stalks finely hairy. Leaves shiny above, minutely hairy below, 1–4.5 cm long, compound with 3 leaflets; 2 side leaflets almost stalkless, terminal leaflet on short stalk. Fruit round, small, to 1 cm in diameter, deep red to dark purple. Inhabits moist to wet open areas such as edges of watercourses, wet meadows, tundra and bogs in northern Ontario. Also called: nagoonberry, northern dwarf raspberry • syn. *R. acaulis.*

Black raspberry (*R. occidentalis*) has long, arching, purple-brown, waxy cane-like stems 1–2 m long, arising in clumps, with hooked prickles. Leaves with 3 or 5 leaflets, often lobed if trifoliate, smooth and green on the upper surface, whitish and fuzzy on the lower. Flowers white, most often in small clusters. Fruit initially bright red then turns black, to 1.5 cm in width at maturity. Restricted to southern Ontario in thickets, ravines and open woods. Also called: eastern black raspberry.

Dwarf raspberry (*R. pubescens*) is a slender, trailing, soft-hairy shrub, unarmed, to 50 cm tall and often more than 1 m long (vegetative stems ascend at first, then recline), rooting where the nodes touch the ground. Leaflets

Arctic dwarf raspberry (*R. arcticus*)

alternate, long-stalked, greenish, smooth or slightly hairy, with 3–5 lobes. Flowers 5-lobed, 1–3 on erect shoots, 15–50 cm, blossoms white, rarely pinkish. Fruits dark red druplets, to 1 cm, smooth, several, not easily separating from spongy receptacle. Found on damp slopes, rocky shores and low thickets as well as deciduous, conifer and mixed woods across the province. Also called: trailing raspberry, dewberry.

Black raspberry (*R. occidentalis*)

Dwarf raspberry (*R. pubescens*)

Wild red raspberry (*R. idaeus*)

Flowering raspberry (*R. odoratus*) is a widely branched, straggling shrub growing 1–2 m tall. Stems covered in reddish, glandular bristly hairs, lacks prickles. Bark reddish-brown, freely shredding. Leaves simple, soft, alternate, 10–20 cm long and wide, 3–5 lobed, maple leaf–like, long-stalked, usually soft-hairy above and below, margins irregularly or sharply toothed. Flowers geranium-like, rose-purple, 3.5–5 cm in diameter, several loosely clustered at branch tips. Fruit 1 cm in diameter, dry, seedy, insipid-tasting, dome-shaped, somewhat flattened. Inhabits shaded, moist forest edges and roadsides, thickets and ravines, commonly found from Lake Erie to the Ottawa Valley and locally elsewhere in the province.

Wild red raspberry (*R. idaeus*) is an erect shrub, to 2 m tall, growing in thickets as it spreads by underground rhizomes. Compound leaves usually of 3, but sometimes 5 and rarely 7, dark green leaflets, smooth above and grey to white and hairy underneath. Fruit bright red, virtually identical to the domesticated raspberry, but smaller. Grows in thickets, open woods, fields and on rocky hillsides across Ontario, less commonly in the north. Also called: American red raspberry • syn. *R. strigosis*.

Wild red raspberry (*R. idaeus*)

Wild Berry Dressing

Wild red raspberry (*R. idaeus*)

Makes about 2 cups

This dressing keeps well in the fridge for up to 10 days.

1 cup mixed tangy wild berries such as raspberries, thimbleberries or blackberries
½ cup olive oil • ¼ cup apple cider vinegar
1 tsp sugar • 2 cloves crushed garlic • 1 tsp salt

Crush the berries, then mix with all the remaining ingredients in a small jam jar. Screw on the lid tightly and shake vigorously.

COMPARE: *Raspberries and relatives (salmonberry, thimbleberry and cloudberry) are all closely related members of the genus Rubus. The best way to distinguish these berries from blackberries is by looking at their fruits: if they are hollow like a thimble, they are raspberries (or relatives), and if they have a solid core, they are blackberries. Before and after fruiting, the plants can be differentiated according to leaf type on non-flowering branches; palmate with 5 leaflets for blackberries and trifoliate for raspberries.*

Wild red raspberry (*R. idaeus*)

73

Dewberries *Rubus* spp.

Northern dewberry (*R. flagellaris*)

Dewberries are delicious, with a deep rich flavour similar to a blackberry. Although the fruit is relatively small, it can grow in profusion and is well worth gathering. Dewberries make excellent jams, jellies, preserves and wine and are also good in baking. The Malecite, Iroquois and Ojibwa ate the fruit fresh or dried.

Take care when gathering these fruit as the prickles are not only extremely sharp but also are brittle and easily break off, leaving them embedded in the skin. They are fine and difficult to remove.

EDIBILITY: highly edible

FRUIT: Fruits drupelets, red or black.

SEASON: Flowers in June. Fruits ripen July to August.

DESCRIPTION: Perennial trailing shrub from a woody taproot with whip-like branches growing to 4.5 m, rooting at the tip. Stems brown to reddish, with extremely sharp, hooked prickles. Leaves green, thin, compound, alternate, mostly hairless, with 3–5 leaflets. Leaves of flowering canes normally smaller, with 3 leaflets. Flowers white, on nearly erect stems growing from nodes of main branches, with 5 petals, solitary or in terminal clusters of 2–5. The fruits may resemble blackberries and other raspberries but are largely differentiated by their trailing growth habit.

Northern dewberry (*R. flagellaris*) grows to 4.5 m in length with deciduous leaves 6–18 cm. Flowers 10–15 mm wide, 1–5 at stem tip. Red to black fruits are attached to the receptacle (core). Inhabits open, dry, rocky or sandy woods and thickets around the Great Lakes of southern Ontario. The specific epithet *flagellaris* derives from the Latin for "whip-like," in reference to the long stems.

Northern dewberry (*R. flagellaris*)

Swamp dewberry (*R. hispidus*) rarely reaches 1 m in height with leaves that often persist through the winter. Fruit is red, ripening to glossy purplish-black about 10 mm across, resembles a small blackberry but differs in that it sometimes separates cleanly from the core, leaving a hollow depression like a raspberry. Inhabits moist, rich forests from northern boreal to southern mixed hardwood as well as swamps, bogs and rocky or disturbed areas particularly after a fire. Also called: dewberry, bristly dewberry.

Swamp dewberry (*R. hispidus*)

Swamp dewberry (*R. hispidus*)

Cloudberry *Rubus chamaemorus*

Also called: bake-apple

Cloudberry (*R. chamaemorus*)

Cloudberry was historically, and is still, a principal food for northern indigenous peoples; the juicy berries are delicious, with a distinctively tart taste that some reports say is acquired. These berries have twice as much vitamin C per volume as an orange and were an important food against scurvy for First Nations and early northern immigrants. Traditionally, these summer fruits were stored in seal pokes, wooden barrels or underground caches in cold water or oil, with other berries or with edible greens.

The Latin term for this species derives from the Greek words "*chamai*," meaning "on the ground," and "*moros*," meaning "mulberry."

EDIBILITY: highly edible

FRUIT: Fruit raspberry-like in appearance, each made up of 5–25 drupelets, amber to yellow when mature.

Cloudberry (*R. chamaemorus*)

Cloudberry (*R. chamaemorus*)

SEASON: Flowers May to June. Fruits ripen in August.

DESCRIPTION: A low, unbranched herb, to 25 cm tall, with 1–3 leaves per stem. Leaves alternate, round to kidney-shaped (not divided into leaflets), shallowly 5- to 7-lobed, no prickles or bristles. Flowers single, white, at end of stem; the male and female flowers on different plants. Found in peat bogs and peaty forests at northern latitudes.

Cloudberry (*R. chamaemorus*)

Berry Fruit Leather

Makes 1 baking sheet of fruit leather

4 cups crushed berries (all one kind or a mix) • 2 cups apple sauce • ½ cup sugar

Mix the berries and sugar together in a pot on medium heat until the sugar is dissolved. Put the mixture through a food mill to remove any stems or seeds, then add the apple sauce and stir until well mixed. Grease a rimmed baking sheet and pour mixture in. Use a spatula to spread the mixture to an even thickness on the baking sheet, because the fruit leather will not dry evenly otherwise. Place in a food dehydrator or an oven at 150° F until firm to the touch and dry enough to peel off. Remove from the dehydrator or oven and let cool. Use scissors to cut the leather into strips. Cool the strips and store in an airtight container or Ziploc® bag.

Thimbleberry *Rubus parviflorus*

Thimbleberry (*R. parviflorus*)

Thimbleberry is one of the most delicious native berries you will encounter and was highly regarded by all First Nations within its range. The fruit is easy to pick as it can grow in large clusters and appears on the plant as bright red treasures amid soft, maple leaf–like leaves (yay, no sharp prickles or spines!). The taste is somewhat like a raspberry, but more intense and flavourful with a sharper "tang." Once you've had thimbleberry pie, jam or tarts, you will never go back to its poorer cousin—the raspberry! The fruit, which is rather coarse and not overly juicy, dries and keeps well. This species can also be gathered by cutting the stems of the unripe fruit, which will ripen later in storage. Traditionally, these fruit were gathered, mashed either alone or with other seasonal

fruits, and dried into cakes for winter use or trade. The tender shoots of this plant were traditionally harvested and peeled in the early spring as a green vegetable.

The large leaves of thimbleberry served many purposes for some indigenous peoples. They were used to whip soapberries, wipe the slime from fish, line and cover berry baskets and dry other kinds of berries. To make a temporary berry container, pick a leaf and then snap off the stem. Fold the outer soft leaf edges together, to form a funnel shape (the stem is at the narrow, bottom edge of this funnel, the leaf tips forming the wider top brim), then use the stem to prick through the two leaf folds where they overlap and "sew" the funnel together. If you still have a small hole at the bottom of your funnel, line this with part of another leaf. If you're out in the woods and have forgotten your toilet paper, thimbleberry leaves are soft and tough and make an excellent substitute.

EDIBILITY: highly edible

FRUIT: Bright red, shallowly domed (like a thimble), raspberry-like hairy drupelets, in clusters held above the leaves.

SEASON: Flowers April to May. Fruits ripen July to August.

DESCRIPTION: Erect shrub, 0.5–3 m tall, main-stemmed with many branches and no prickles or spines, spreading by underground rhizomes and forming dense thickets. Bark light brown and shredding on mature stems, green on newer stems. Leaves large (up to 15 cm), soft, fuzzy, maple leaf–like, palmate, 3–7-lobed, alternate, toothed around margins, fine hairs above and below. Flowers white, 5-petalled, large, to 4 cm, long-stemmed in clusters of 3–11, terminally clustered. Found in moist open sites such as road edges, shorelines, riverbanks and open forests along Lake Superior and Huron to the Bruce Peninsula.

Thimbleberry (*R. parviflorus*)

Thimbleberry (*R. parviflorus*)

Wild Berry Juice

Makes approximately 4½ cups

3 cups any sweet berries such as blueberries, thimbleberries or blackberries

2 cups water • sugar to taste

Pick over berries to remove any debris and place them in a saucepan with the water. Mash the mixture with a wooden spoon on potato masher, then simmer until berries are soft. Strain the mixture through a jellybag, fine-mesh sieve or cheesecloth, add sugar to taste, then let the juice cool before serving.

Mulberries *Morus* spp.

Red mulberry (*M. rubra*)

Juicy, red mulberries are an explosion of dark, fruity taste somewhat like a mix of a perfectly ripe blackberry and blackcurrant. The berries are notorious for creating hard-to-remove stains on clothing and pavement, but the taste of the fruit is well worth the possible trouble! The Iroquois and Huron peoples enjoyed the berries of red mulberry fresh or preserved but recognized that the unripe berries and milky sap were toxic and could be irritating to the skin.

Native American tribes used mulberry leaves as a tea to treat weakness and urinary problems and the sap to treat ringworm. White mulberry has a long history of medicinal use in China, where the leaves are considered to have antibacterial, astringent, hypoglycemic and ophthalmic properties. Leaves are gathered in autumn after the first frost and are used either fresh or dried. The stems are gathered in late spring or early summer and are used as a diuretic and to treat rheumatism. The leaves of white mulberry are the main source of food for silkworms in eastern Asia. The milky juice is rich in rubber-like compounds said to add tenacity to the silk fibres spun by the worms. The white mulberry has been cultivated for thousands of years in China as part of the silk industry, where the trees are pollarded (cut back to the trunk) to stimulate a dense head of leafy shoots to feed the silkworms. As long ago as the 3rd century bc, Chinese silk caught the eye of the Romans, who encountered it on the "silk roads" over which Chinese goods were traded to the West. In later centuries, various forces (e.g., Arab conquerors in Persia, Crusaders and Marco Polo's journeys to China) facilitated the growth of a silk industry in European countries such as Italy, France and Germany. Silk production flourished there

White mulberry (*M. alba*)

for several hundred years, until World War II interrupted the flow of raw silk from Japan and effectively ended silk-making in Europe. White mulberry shrubs, along with silkworms, were brought to North America in the 1800s in an attempt to establish a Western silk industry. This ambitious undertaking soon failed, but since then, these hardy trees have thrived and spread across southern Ontario.

Mulberry is an attractive, fast-growing species that is planted as an ornamental or fruit tree, but it requires sufficient space to accommodate its spreading branches. Female trees attract many birds and small mammals to parks and yards and are planted to lure birds away from other fruit. Mulberry wood is very durable and has been used in ornamental carving and for making fences, barrels and boats.

EDIBILITY: highly edible

FRUIT: Fruits rounded, blackberry-like; clusters (multiple fruits) 1–3 cm long, composed of tiny, seed-like fruits (achenes), each surrounded by a small, juicy segment.

SEASON: Flowers April to May. Fruits ripen June to August.

DESCRIPTION: Deciduous shrubs or small trees, 10–20 m tall, with milky sap. Leaves alternate, simple, toothed and oval to heart-shaped, 10–20 cm long, yellow in autumn. Leaves can also be deeply lobed. Flowers tiny, green; male and female flowers in separate clusters. Male clusters loose and elongated; female clusters short, dense and cylindrical; appear in early spring.

Red mulberry (*M. rubra*) generally grows to 10–15 m in height with dark brown bark either smooth or scaly. Leaves longer than wide, hairy beneath with finely serrated edges and egg-shaped, often lobed. The fruits are red to dark purple or almost black, up to 3 cm long. Native to southern Ontario, where it grows in moist, rich forests and floodplains.

White mulberry (*M. alba*) has grey or light brown bark, initially smooth but forming narrow scaly ridges with age. Leaves are hairless (or with tufts of hairs underneath) and variable in shape, from ovate to intricately lobed. White, red or purple to blackish fruits. Native to China, but naturalized in open, disturbed sites, along fences and near the edges of forests and in planted gardens in southern Ontario.

Red mulberry (*M. rubra*)

White mulberry (*M. alba*)

Barberries   *Berberis* spp.

Japanese barberry (*B. thunbergii*)

Though not native to North America, these dense, viciously thorny shrubs provide edible and medicinal benefits that have a long history of use. Barberry fruits are acceptable raw in small quantities, but they are very acidic and instead make excellent preserves or jellies. In Iran, dried Berberis berries, called zereshk, are widely used and impart a tart flavour to chicken and rice dishes. A popular candy in the Ukraine of the same name is also made from the berries. A rich source of vitamin C and the soluble fibre pectin, barberry berries make a refreshing lemonade-like cold beverage when sweetened.

Barberry contains a number of well-studied alkaloids, especially berberine (found mostly in the roots), which is antimicrobial, anti-inflammatory and astringent. It has been tested for its potential usefulness in treating diabetes, prostate cancer, cardiac arrhythmia and leukemia, although not enough research has been conducted in humans. By itself, berberine is a relatively weak antibiotic. However, an extract that contains barberry's other components, such as the isoquinoline alkaloids berbamine and oxyacanthine, displays significantly stronger antibacterial activity, as well as effects against amoebas and trypanosomes. These alkaloids are poorly absorbed through the digestive tract, so they are particularly useful against enteric infections, such as bacterial dysentery, and parasite infections. In this regard, herbalists do not recommend that wild licorice be consumed at the same time because it is surmised that it nullifies the effects of barberry. The clinical use of purified berberine today is mainly

Common barberry (*B. vulgaris*)

to counteract bacterial diarrhea, and in eye drops to treat ocular trachoma infections, hypersensitive eyes, inflamed eyelids and conjunctivitis. Berberine can be poisonous if taken in large doses (see Warning).

A mild laxative can be made from an infusion of the berries mixed with wine. A decoction of the berries or root bark makes an effective mouthwash or gargle for mouth and throat complaints. Fresh barberry juice was thought to strengthen the gums and relieve pyorrhea when brushed directly on them.

Barberry shrubs are sometimes planted as hedges and as barriers under vulnerable windows to deter trespassers. They tolerate trimming well. Barberry is the intermediate host for wheat rust and was the focus of an extensive eradication effort that lasted from 1918 to 1975.

EDIBILITY: edible (with caution)

FRUIT: Fruits a small berry, elliptical and scarlet when mature, usually long and narrow, resembling a bar (hence the common name barberry).

SEASON: Flowers from May to June. Berries ripen August to September.

DESCRIPTION: Perennial, deciduous shrubs growing 0.5–3 m tall. Branches grey. Long shoots erect, branched or unbranched, with simple, 3 or 5-spined thorns 3–30 mm long. Short shoots in the thorn axils produce leaves to 10 cm long, margins entire or spiny. Leaves 2.5–7.5 cm long and bristle-toothed. Flowers produced in pendulous clusters of 10–20 per head, yellow, 3–6 mm long, with 6 petals and sepals.

Common barberry (*B. vulgaris*) can grow to 3 m in height, smooth grey branches bearing sharp 3-pronged spines, often with leafy shoots growing from the axils. Leaves hairy on both sides, widest above the middle and spiny-toothed. Bright scarlet, oblong berries hang in clusters. This introduced Eurasian species is widely naturalized in disturbed spaces (pastures, roadside thickets) in southern Ontario. Also called: European barberry.

Japanese barberry (*B. thunbergii*) is a compact shrub that reaches 30 cm–3 m high with purple or brown stems, single spines or in clusters of 3–5 and smooth-edged leaves usually less than 3 cm long. Berries red, elliptical to round, persist through winter. This introduced ornamental grows in open woodlands, fields and roadsides of southern Ontario.

WARNING: *Berberine is toxic at high doses and reportedly disrupts proper liver function in infants. Barberries should not be consumed during pregnancy and breast-feeding. Strong barberry preparations may lead to intestinal discomfort or adverse interactions with the antibiotic tetracycline.*

Common barberry (*B. vulgaris*)

Tall Oregon-grape   *Mahonia aquifolium*

Also called: Oregon grape, holly-leaved barberry • *Berberis aquifolium*

Tall Oregon-grape (*M. aquifolium*)

This berry is the only Oregon-grape species that occurs in Ontario. Like other species to the west, these tart, juicy berries can be eaten raw but can be rather sour and intense, so are more commonly used to make jelly, jam or wine. A frost increases the fructose content of the berries, thereby making them sweeter and more palatable for fresh eating. Mashed with sugar and served with milk or cream, they make a tasty dessert. A refreshing drink can be made with mashed berries, sugar and water—the sweetened juice tastes much like concord grape juice. Berry production can vary greatly from year to year, and the fruits are sometimes rendered inedible by grub infestations, so the eater should be wary of the potential for extra protein!

The plants of this species contain the alkaloids berberine, berbamine, isoco-rydin and oxyacanthine, which stimulate involuntary muscles. The crushed plants and roots have antioxidant, antiseptic and antibacterial properties. They were used to make medicinal teas, poultices and powders for treating gonorrhea and syphilis and for healing wounds and scorpion stings. Boiled, shredded root bark produces a beautiful, brilliant yellow dye.

Tall Oregon-grape is a decorative, tough, spiky, drought-tolerant and hardy plant that is often planted as an ornamental. It produces masses of bright yellow, scented flowers adored by pollinators early in the spring, followed by tresses of decorative purple berries loved by birds, and eventually a striking and lasting fall display of red leaves. Its sharp spiky leaves also make it a good plant in areas where you may want to discourage pets or people from exploring.

EDIBILITY: edible (with caution)

FRUIT: Fruits juicy, grape-like berries, tart and sour tasting, about 1 cm long, purplish blue with a whitish bloom.

SEASON: Flowers April to June. Fruits ripen August to September.

DESCRIPTION: Perennial, evergreen shrub to 4 m tall. Outer bark rough, brown to greyish, and with a striking canary/mustard yellow inner layer when scraped. Leaves leathery, holly-

Tall Oregon-grape (*M. aquifolium*)

WARNING: *Like closely related barberries, Tall Oregon-grapes contain alkaloids such as berberine and should not be consumed during pregnancy or breast-feeding. High doses can irritate the skin, nose and eyes and cause diarrhea, vomiting, and liver and kidney problems.*

like, pinnately divided into 5–11 spiny-edged leaflets, dark glossy green with a prominent central vein on each, turning red or purple in winter. Flowers yellow, about 1 cm across, in upright or in hanging, whorled clusters. Grows escaped in dry forests open fields in southern Ontario.

Blackberry and Oregon-grape Jelly

Makes 16 x 1 cup jars

8 cups blackberries • 8 cups Oregon-grape berries • ¼ cup lemon juice
1 packet powdered pectin • 5 cups sugar

Place blackberries and Oregon-grape berries in a thick-bottomed saucepan on medium heat. Crush and stir the berries and simmer until the juice is released, about 10 minutes. Strain through a cheesecloth or fine-mesh sieve to separate the juice. *Do not squeeze the cloth or force the mix through the sieve because it will cause sediments to run into the juice, resulting in a cloudy jelly.*

Measure out 4 cups of the resulting juice into a thick-bottomed saucepan. Add the lemon juice and pectin, stirring until the pectin is thoroughly dissolved. Add the sugar, stirring constantly, and bring to a rolling boil. Hold the mixture at the boil for 3 minutes, being careful to stir the bottom so that the jelly does not stick or burn.

Meanwhile, prepare 16 x 236 mL jars and lids (wash and sterilize jars and lids, and fill jars with boiling water; drain just before use).

Remove from heat, skim off any foam (the impurities coming out of the liquid) and pour into the hot, sterilized jars. Carefully wipe the jar edges to ensure they are clean and dry, then place the lids on and tighten the metal screw bands. Place jars in a cool area. You will know that the jars have sealed when you hear the snap lids go "pop."

Sassafras   *Sassafras albidum*

Sassafras (*S. albidum*)

The fruit of this species is variably reported as tasty to toxic and poisonous so is not recommended for consumption. The Iroquois nations in Canada and the United States used sassafras twigs as chewing sticks, the leaves as seasoning in meat soups and the bark and roots as a spice. Sassafras was also quite popular among most First Peoples in its range as a tea made from the leaves, roots and/or flowers. The leaves can be eaten raw (added to salads) or cooked; they have a mild, aromatic flavour. The young shoots have been used to make beer. The root has been prepared in a number of ways, such as brewing with maple syrup, to be used as a condiment, tea or flavouring. Sassafras root was one of the original ingredients of rootbeer; however, the essential oil is now banned as a food flavouring in America.

Sassafras was widely used as a medicine by indigenous peoples and was especially valued for its tonic effects. Parts of the sassafras plant (roots, pith from new sprouts, bark) were often taken for blood ailments: for example, high blood pressure, to thin the blood, for nosebleeds, as a blood purifier, to treat watery blood or to "clear" the blood. A compound infusion of sassafras roots and whisky was taken for tapeworms and rheumatism, and the leaves were applied as a poultice for wounds, cuts and bruises. It was often used to treat female ailments such as painful menstruation and fevers after childbirth. The roots were also applied to treat swellings on the shins and calves, and a wash was prepared from the whole plant to treat sore eyes or cataracts. As a folk medicine, sassafras tea made from root bark was renowned as a spring tonic and blood purifier; it was considered a household cure for gastrointestinal complaints, colds, kidney problems, rheumatism and skin eruptions. Sassafras root was occasionally used in commercial dental poultices. Sassafras oil has been applied externally to control lice and treat insect bites, but it can cause skin irritations. The oil has also been used in soaps, perfumery, toothpastes and soft drinks. A yellow dye can be derived from the wood and bark.

EDIBILITY: edible with caution (toxic), poisonous

FRUIT: Fruit a dark blue, ovoid to round single-seeded "berry" 1 cm across, on a red stalk.

SEASON: Flowers April to May. Fruits ripen in September.

DESCRIPTION: Medium to large, deciduous tree, to 35 m tall, more bush-like in some conditions. Bark dark red-brown, deeply furrowed. Twigs pale green with darker olive mottling. Leaf blade to 16 cm long, oval to elliptic, either unlobed or with 2–3 lobes with characteristic aroma when crushed. Flowers greenish yellow, fragrant (sweet lemony), to 5 cm across, borne in clusters. Found in disturbed areas, forests, woodlands and old fields south of Toronto.

> **WARNING:** Safrole *is a main component of sassafras oil (80 percent). In large doses, it can cause dilated pupils, vomiting, kidney and liver damage and is a suspected carcinogen.* Safrole *may also cause dermatitis.*

Sassafras (*S. albidum*)

Sassafras (*S. albidum*)

87

Currants *Ribes* spp.

Red swamp currant (*R. triste*)

Common and widespread throughout Ontario, currants were eaten by many First Nations and colonial settlers. All are considered edible, some are tastier than others, and some are considered emetic in large quantities and are best avoided. Currants are high in pectin and make excellent jams and jellies, either alone or mixed with other fruit. These preserves are delicious with meat, fish, bannock or toast.

Historically, currants were also combined with other berries and used to flavour liqueurs or fermented to make delicious wines, but raw currants tend to be very tart.

Golden currant is one of the most flavourful and pleasant-tasting currants. Skunk currants have a skunky smell and flavour when raw but are delicious cooked.

Some First Nations used skunk currant stems to prevent blood clotting after childbirth, American currant roots for uterine problems and kidney ailments, and the branches of northern black currant for colds, stomach problems and sore throats. In Europe, currant juice is taken as a natural remedy for arthritic pain. Black currant seeds contain gamma-linoleic acid, a fatty acid that has been used in the treatment of migraine headaches, menstrual problems, diabetes, alcoholism, arthritis and eczema.

Some Native peoples believed that northern black currant had a calming effect on children, so sprigs were often hung on baby carriers. Currant shrubs growing by lakes were seen as indicators of fish; in some legends, when currants dropped into the water, they were transformed into fish.

For more information on closely related species, see gooseberries and prickly currants.

EDIBILITY: edible

Golden currant (*R. aureum*)

Northern black currant (*R. hudsonianum*)

FRUIT: Fruit colour varies from bright red to green to black, as do sweetness and juiciness, depending on the species and individual location.

SEASON: Flowers April to July. Fruits ripen July to August.

DESCRIPTION: Erect to ascending, deciduous shrubs, 1–3 m tall, without prickles, but often dotted with yellow, crystalline resin-glands that have a sweet, tomcat odour. Leaves alternate, 3- to 5-lobed, usually rather maple leaf–like. Flowers small (about 5–10 mm across), with 5 petals and 5 sepals, borne in elongating clusters in spring. Fruits tart, juicy berries (currants), often speckled with yellow, resinous dots or bristling with stalked glands.

Golden currant (*R. aureum*) grows 1–3 m tall and is named for its showy, bright yellow flowers, not its smooth fruits, which range from black to red and sometimes yellow. Its leaves have 3 widely spreading lobes and few or no glands. Inhabiting streambanks and wet grasslands to dry prairies and open or wooded slopes, it is uncommon in southern Ontario.

Northern black currant (*R. hudsonianum*) grows to 1 m tall, with elongated clusters of 6–12 saucer-shaped, white flowers or shiny, resin-dotted, black berries. Its relatively large, maple leaf–like leaves have resin dots on the lower surface. Fruit strong-smelling and often bitter-tasting. Found in wet woods, swamps and thickets or on rocky slopes in western, northern and parts of southern Ontario. Also called: Hudson Bay currant.

Red swamp currant (*R. triste*) is an unarmed, reclining to ascending

Northern black currant (*R. hudsonianum*)

shrub, rarely reaching a height of 1.5 m. Leaves to 10 cm in width and length, 3–5 lobed and paler, usually hairy undersides. Flowers reddish or greenish purple, small, several (6–15) in drooping clusters; flower stalks jointed, often hairy and glandular.

Golden currant (*R. aureum*)

Red swamp currant (*R. triste*)

Fruits bright red, smooth and sour but palatable. Found in moist, coniferous forests, swamps, on streambanks and montane, rocky slopes. Also called: wild red currant.

Skunk currant (*R. glandulosum*) is a loosely branched, unarmed, trailing shrub 0.5–1 m tall with spreading stems. Bark is brownish to purple-grey. Leaves 3–8 cm wide and similarly long, 5–7 lobed, glabrous. Flower stalks not jointed below the flowers. Berries dark red, 6–8 mm, nearly

round in shape, bristly and considered not very good to eat. Found in cool, moist and rocky woods and along streams and shores across the province.

Wild black currant (*R. americanum*) is a small, non-prickly shrub growing to 1 m or more in height. Leaves simple, rounded and alternate, to 10 cm wide and 8 cm long, palmately lobed, with 3–5 pointed lobes with double-toothed edges. The surfaces of the leaves are scattered with resinous dots. Flowers creamy white to yellowish, bell-shaped and hanging in clusters from the leaf axils. Fruit a black berry 6–9 mm, globular, smooth, each with a characteristic residual flower at the end. Found in damp soil along streams, wooded slopes, open meadows and rocky ground, common in southern and eastern Ontario and locally up to 50º N. Also called: American black currant.

Skunk currant (*R. glandulosum*)

Wild black currant (*R. americanum*)

Gooseberries *Ribes* spp.

Northern gooseberry (*R. oxyacanthoides*)

While gooseberries and currants are closely related species, they are generally different in that gooseberries have spines or prickles on their stems (currants are not thus "armed") and gooseberry fruit are usually borne in small clusters or singly (currants are in elongated clusters generally more than 5). However, common names are inconsistent so some "gooseberries" don't have spines and some "currants" do!

All gooseberries are edible raw, cooked or dried, but flavour and sweetness vary greatly with species, habitat and season. They are high in pectin and make excellent jams and jellies, either alone or mixed with other fruits. Gooseberries can be eaten fresh and are also used in baked goods such as pies. Traditionally, these fruits were eaten with grease or oil, and also mashed (usually with other berries) and formed into cakes that were dried and stored for winter use. Dried gooseberries were sometimes included in pemmican. Because of their tart flavour, gooseberries can be used much like cranberries. They make a delicious addition to turkey stuffing, muffins and breads. Timing is important, however, when picking these fruit. Green berries are too sour to eat, and ripe fruit soon drops from the branch. Sometimes green berries can be collected and then stored so that they ripen off the bush. Consuming too many gooseberries can cause stomach upset, especially in the uninitiated.

Because of the large number of species and wide distribution of gooseberries, there is a very large spectrum of uses for this genus. They were commonly eaten or used in teas for treating colds and sore throats, a practice that may be related to their high vitamin C content. Teas made from gooseberry leaves and fruits were given to women whose uteruses had slipped out of place after too many pregnancies. Gooseberry tea was also used as a wash for soothing skin irritations such as poison-ivy rashes and erysipelas (a condition with localized inflammation and fever caused by a Streptococcus infection). Gooseberries have strong antiseptic properties, and extracts have proved effective against yeast (Candida) infections. Picking this fruit can be a formidable task, though, because of the often thorny stems. Indeed, gooseberry thorns can be so large and strong that they were historically used as needles for probing boils, removing splinters and even applying tattoos! The name "gooseberry" comes from an

Northern gooseberry (*R. oxyacanthoides*)

old English tradition of stuffing a roast goose with the berries.

EDIBILITY: edible

FRUIT: Fruits smooth, purplish (when ripe) berries, about 1 cm across.

SEASON: Flowers May to June. Fruits ripen July to August.

DESCRIPTION: Erect to sprawling deciduous shrubs generally less than 1.5 m high with spiny branches. Leaves alternate, maple leaf–like, 3- to 5-lobed, 2.5–6 cm wide and coarsely toothed. Flowers whitish to pale greenish yellow, to 1 cm long, tubular or bell-shaped, with 5 small, erect petals and 5 larger, spreading sepals, in 1- to 4-flowered inflorescences in leaf axils.

Northern gooseberry (*R. oxyacanthoides*)

globose, greenish to pale or purplish red, 7–15 mm, densely spiny or bristly. Common in southern Ontario to just north of Lake Huron, inhabits rocky slopes, heaths, rich hardwood and conifer forests. Also called: dogberry.

Northern gooseberry (*R. oxyacanthoides*) is an erect to sprawling shrub, 1–1.5 m tall, branches bristly, often with 1–3 spines to 1 cm long at nodes; internodal bristles absent or few. Twigs grey to straw-coloured, older bark whitish grey, peels less readily and so remains prickly longer than wild gooseberry. Smooth, blue-black berries up to 12 mm across. Inhabits wet forests, thickets, clearings, open woods and exposed rocky sites of northern and western Ontario. Also called: smooth gooseberry, bristly wild gooseberry • syn. *R. setosum*.

Prickly gooseberry (*R. cynosbati*) is a low shrub with stems smooth or covered with fine soft hairs, internodal prickles sparse or absent, to 1 cm. Leaves 2–7 cm long and wide, surface smooth or soft-hairy, margins with rounded teeth. Bell-shaped flowers solitary or 2–4, greenish-white. Berries

Wild gooseberry (*R. hirtellum*)

Wild gooseberry (*R. hirtellum*) has ascending or erect branches and greyish branchlets armed with scattered prickles and 1 to 3 sharp, 3–8 mmlong spines. Outer bark peels with age revealing the dark reddish to black inner bark. Leaves 3–5 lobed, to 6 cm long and wide, dark green above, and lighter and hairy underneath. Greenish-yellow to purplish, narrowly bellshaped flowers, 2–3 together from leaf axils. Berries are smooth and round, bluish-black and 8–12 mm in diameter. Found in bogs, lowland valleys and on streambanks and open or wooded, montane slopes, locally in the north and more commonly elsewhere. Some taxonomists consider these plants a variety of *R. oxyacanthoides*.

Northern gooseberry (*R. oxyacanthoides*)

Northern gooseberry (*R. oxyacanthoides*)

Prickly Currant *Ribes lacustre*

Also called: swamp gooseberry, bristly black currant, swamp blackcurrant

Prickly currant (*R. lacustre*)

Despite the prickly and strongly irritating stems of this plant, the fruit is quite palatable when ripe and makes delicious jam and pies (either alone or mixed with other fruit) for those who survive the spikes. This is an unusual species in that it has characteristics common to both currants and gooseberries. Like a gooseberry, prickly currant is covered in many sharp, spiny prickles that are also highly irritating. Like a currant, its fruit hang in clusters.

Prickly currants were widely used for food by First Nations, although their flavour is sometimes described as insipid. The berries were traditionally eaten fresh off the bush, cooked, stored in the ground for winter use, or sometimes dried. Dried currants were occasionally included in pemmican, and

they make a tasty addition to bannock, muffins and breads. Currants may be boiled to make tea or mashed in water and fermented with sugar to make wine. The leaves, branches and inner bark of prickly currant produce a menthol-flavoured tea, sometimes called "catnip tea." To make this tea, the spines were singed off, and the branches (fresh or dried) were steeped in hot water or boiled for a few minutes. Prickly currant fruits were traditionally rolled on hot ashes to singe off their soft spines before eating.

Some Native peoples considered the spiny branches (and by extension, the fruit) of this plant to be poisonous. However, this shrub was also useful as its thorny branches were thought to ward off evil forces.

Prickly currant (*R. lacustre*)

Prickly currant (*R. lacustre*)

EDIBILITY: edible

FRUIT: Shiny, dark purple berry, hanging in drooping clusters and covered in long hairs.

SEASON: Flowers in April. Fruits ripen June to August.

DESCRIPTION: Erect to spreading deciduous shrub, 50 cm–1 m tall, covered with numerous small, sharp prickles, and larger, thick thorns at leaf nodes. Bark on older stems is cinnamon-coloured or pale brown. Leaves large (4–8 cm wide and long) and hairless to slightly hairy, alternate, 3- to 5-lobed, maple leaf–like. Flowers reddish to maroon, saucer-shaped, about 6 mm wide, forming hanging clusters of 7–15. Fruits 5–8 mm-wide berries, bristly with glandular hairs. Inhabits moist woods and streambanks to drier forested slopes and subalpine ridges. Present throughout the province except the far west and southern deciduous forests.

Prickly currant (*R. lacustre*)

Serviceberries   *Amelanchier* spp.

Also called: Saskatoon, Canada serviceberry, juneberry, shadbush

Saskatoon berry (*A. alnifolia*)

The berries of these plants were and still are extremely valuable to many indigenous peoples across Canada. Indeed, there is a well-documented history of extensive landscape management through fire, weeding and pruning to encourage the healthy growth of these important plants. Large quantities of the berries were harvested and stored for consumption during winter. They were eaten fresh, dried like raisins, or mashed and dried into cakes for winter use or trade. Some indigenous peoples steamed serviceberries in spruce-bark vats filled with alternating layers of red-hot stones and fruit. The cooked fruit was mashed, formed into cakes, then dried over a slow-burning fire. These cakes could weigh as much as 7 kg (15 lbs) each!

Dried serviceberries were the principal berries mixed with meat and fat to make pemmican and were commonly added to soups and stews. Today, they are popular in pies, pancakes, puddings, muffins, jams, jellies, sauces, syrups and wine, much like blueberries. With a number of strong polyphenolic antioxidants, serviceberries are of increasing agricultural and economic interest as a health food.

Historically, serviceberry juice was taken to relieve stomach upset and was also boiled to make drops to treat earache while green or dried berries were used to make eye drops. The fruits were given to mothers after childbirth for afterpains and were also prescribed as a blood remedy. The berry juice, which easily stains your hands when picking, makes a good purple-coloured dye.

Serviceberries are excellent ornamental, culinary and wildlife shrubs. They are hardy and easily propagated, with beautiful white blossoms in spring, delicious fruit in summer and colourful, often scarlet leaves in autumn. There are a number of improved garden cultivars available for superior fruit production in the home garden.

EDIBILITY: highly edible

FRUIT: Fruits juicy, berry-like pomes, red at first, ripening to purple or black, sometimes with a whitish bloom, 6–12 mm across.

SEASON: Flowers April to June. Fruits ripen June to August.

DESCRIPTION: Shrub or small tree, to 10 m tall, often forming thickets and with a massive root system. Bark is smooth, grey to reddish brown. Leaves alternate, coarsely toothed on the upper half, leaf blades 2–5 cm long, oval to nearly round, yellowish orange to reddish brown in autumn. Flowers are white, forming short, leafy clusters near the branch tips; petals are slightly twisted. Grows at low to middle elevations in prairies, thickets, hillsides and dry, rocky shorelines, meadows, open woods.

Saskatoon berry (*A. alnifolia*)

Saskatoon Squares

¼ cup butter • ⅔ cup brown sugar
1 tbsp vanilla • 1 large egg, beaten
1 cup flour • 1 tsp baking powder
½ tsp salt • ½ tsp cinnamon
½ cup FROZEN saskatoons (or wild blueberries)
½ cup chopped walnuts or almonds

Preheat oven to 350° F. Melt the butter gently in a saucepan, then remove from heat and stir in sugar, vanilla and beaten egg. Mix dry ingredients in a bowl. Make a shallow well in the middle, and gradually mix in wet ingredients from the saucepan. When it's well mixed, add frozen saskatoons and chopped nuts. Pour into an 8-inch pan and bake for 35 minutes. Remove from oven and cool before cutting into squares.

WARNING: *The fresh leaves and pits of this species contain poisonous cyanide-like compounds. However, cooking or drying destroys these toxins.*

Pemmican

Makes 6 cups

The pemmican uses the same drying temperature as the fruit leather (see p. 73), so make both recipes at the same time!

3 Tbsp salted butter • 3 Tbsp brown sugar
¼ tsp dried powdered ginger
¼ tsp ground cloves • ¼ tsp ground cinnamon
4 cups saskatoons or blueberries
4 cups beef jerky, chopped into small pieces
½ cup chopped almonds, walnuts or hazelnuts (optional)
½ cup sunflower seeds (optional)

Gently heat butter with sugar and spices in a heavy-bottomed pot. Mash berries and add to pot. Simmer, stirring constantly, for about 5 minutes. Let mixture cool, then mix in jerky and nuts and/or seeds. Grease a rimmed baking sheet, spread mixture evenly on sheet and let dry overnight in oven at 150° F.

Canada serviceberry (*A. canadensis*) has leaf blades 5–9 cm long, oblong, covered with fine, soft, grey fuzz when young, smooth when mature. Flowers with petals to 1.2 cm long, in ascending clusters. Native to southern Ontario, the species is uncommon and found in moist woods, swamps and thickets.

Downy serviceberry (*A. arborea*) grows to 10 m tall and has leaf blades 5–9 cm long and 2–4 cm wide (widest below the middle), hairy beneath, with fine, sharply toothed margins and a pointed tip. Flowers with petals to 2 cm long or dry, red-purple fruits to 12 mm in nodding clusters. Found in southern Ontario in open fields and woods, rocky ridges, forest edges and sandy bluffs. Also called: downy juneberry, shadbush.

Saskatoon berry (*A. alnifolia*)

veins irregularly spaced and above the middle curved toward the tip. Fruits purple to black, sweet and juicy, almost round (6 mm in diameter), on 1–2 cm pedicels. Also called: dwarf or running serviceberry.

Downy serviceberry (*A. arborea*)

Low juneberry (*A. spicata*) grows 0.3–2 m tall, spreading by surface runners to form colonies. Branchlets covered in silky hairs and purplish-red when young. Mature bark is brownish-grey, smooth. Leaves 2–5 cm long, rounded and somewhat pointed at tip, broadly oval to elliptic below, 2–3.5 cm wide, densely wooly on the undersides when young, maturing to hairless. Leaf

Downy serviceberry (*A. arborea*)

Mountain juneberry (*A. bartramiana*) grows to 2 m tall with purplish, mostly hairless branchlets. Leaves to 5 cm long and 2.5 cm wide, on stalks 2–10 mm long, upper surface green and paler beneath, margins often shaded purplish red when unfolding. Distinguished by the close, fine, sharp teeth on the leaves and by the small clusters (1–4) of flowers or pear-shaped pomes 1–1.5 cm long. Inhabits acidic areas wetter than most serviceberries but still grows on drier sites, commonly from central to western Ontario but less so in the south and east.

Red-twigged serviceberry (*A. sanguinea*) grows 1–3 m tall in an erect or straggling growth habit. Branchlets reddish then becoming grey with maturity. Leaf blades 3–7 cm long, 2–4.5 cm wide, heart-shaped to rounded at base, rounded to bluntly pointed at tip. Woolly and folded when young, maturing to hairless. Teeth on leaves, less than twice the number of conspicuous veins. Round, dark purple fruits, 5–10 mm, on 1–3 cm pedicels.

Inhabits gravely and rocky soils or sandy woods and open sites throughout Ontario north to the 51st parallel. Also called: round-leaved serviceberry, shadbush.

Saskatoon berry (*A. alnifolia*) is a bush 1–4 m tall, bark smooth, grey on older stems and red-brown, hairy on new ones. Leaves deciduous, blades 2–5 cm long, oval to nearly round. Flowers with petals to 2.5 cm long. Fruit 6–10 mm, dark blue to black, juicy. Uncommon in western and northwestern Ontario, grows in sandy thickets, edges of woods and streambanks. Also called: serviceberry • syn. *A. humilis*.

Smooth serviceberry (*A. laevis*) grows 5–10 m tall, with smooth and grey trunk bark. Branchlets smooth, purplish. Leaves reddish when younger, up to 8 cm long and 4 cm wide, teeth pointed and sharp on leaf margins, more than twice as many teeth as lateral veins. Fruits dark-red to black on long pedicels (stalks) up to 5 cm long. Inhabits dry loamy or sandy woods and open rocky areas, clearings and thickets in the southern parts of the province.

Mountain juneberry (*A. bartramiana*)

Smooth serviceberry (*A. laevis*)

Dogwoods *Cornus* spp.

Red-osier dogwood (*C. sericea*)

The fruits of this species are rather bitter for modern-day tastes. The berries of red osier dogwood, despite this bitterness, were nonetheless gathered by some First Nations in late summer and autumn and eaten immediately. They were also occasionally stored for winter use, either alone or mashed with sweeter fruits such as serviceberries, and in more modern times with sugar. Red osier dogwood fruit can be cooked when fresh to release the juice, which purportedly makes a refreshing drink when sweetened. Berries of this species with a bluish tinge are said to be the sourest. Some people separated the stones from the mashed flesh and saved them for later use. The stones were then eaten as a snack, somewhat like peanuts are today, but this is not recommended in large quantities and the taste is probably not worth the effort involved.

Some First Nations used the bark for tea to treat digestive disorders because of its laxative effects. The bark infusion was also used to relieve swollen legs and treat venereal disease. The soft inner white bark was smoked to treat lung ailments and for enjoyment. The smoke from the bark was said to be both aromatic and pungent and to have a narcotic effect that could cause stupefaction. Usually, the bark was mixed with tobacco or common bearberry. The inner bark from the roots was included in herb mixtures that were smoked in ceremonies. The inner bark contains an analgesic, coronic acid, which was used as a salicylate-free painkiller.

The flexible branches of red osier dogwood are often used in wreaths or woven into contrasting red rims and designs in baskets. These attractive shrubs, with their lush green leaves and white flowers in spring, red leaves in autumn and bright red branches and white berries in winter, are popular ornamentals with good wildlife and aesthetic values. They grow best in moist sites and are easily propagated from cuttings or by layering.

EDIBILITY: edible, not palatable

FRUIT: Fleshy, berry-like drupe.

SEASON: Flowers May to June. Fruits ripen July to August.

DESCRIPTION: Erect to sprawling, deciduous shrubs or small trees with opposite branches. Leaves opposite, simple, pointed, toothless, with leaf veins following the smooth leaf edges toward the tips, greenish above, white to greyish below, becoming red in autumn. Tiny flowers surrounded by petal-like bracts that resemble a single large flower. Flowers radially symmetrical with 4 sepals, petals and stamens, all attached at the top of the ovary.

Alternate-leaved dogwood (*C. alternifolia*) is a small tree or large scraggly shrub with a short trunk, growing 4–6 m tall with a flat, layered crown. Branches almost horizontal, tips upcurved, in tiers. Leaves alternate, smooth or slightly wavy-edged, with 4–5 parallel side veins arched toward the tip, clustered together at branch tips. White to cream-coloured flowers appear in June. Fruits on short red stalks, maturing July to August, dark blue to bluish-black berry-like drupes 6 mm across. Inhabits moist rich soils in open woodlands and slope bases of southern Ontario. Also called: pagoda dogwood, green osier, blue dogwood.

Red-osier dogwood (*C. sericea*)

Alternate-leaved dogwood (*C. alternifolia*)

Eastern flowering dogwood (*C. florida*)

Eastern flowering dogwood (*C. florida*) is a small tree or tall shrub, to 5 m tall, with a spreading, bushy crown and dark reddish-brown bark. Clusters of small, yellowish flowers, about 30 in a cluster to 1.5 cm wide, subtended by 4 white or pinkish-tinged, showy bracts. Shiny, berry-like red drupes 10–12 mm long in clusters of 3–6.

Eastern flowering dogwood (*C. florida*)

Found in wet or sandy forests and ravines in southeastern Ontario.

Red-osier dogwood (*C. sericea*) is a shrub, to 4 m tall, with bright red twigs and branches (green when young). Leaves to 15 cm long and 9 cm wide, dark green above and pale white beneath. Clusters of small flowers, 2–4 cm wide, without showy bracts, from May to August. Fruits white (sometimes bluish), berry-like drupes, to 9 mm across, containing large, flattened stones. Grows on moist sites, shores and thickets throughout Canada. Also called: syn. *C. stolonifera*.

Round-leaved dogwood (*C. rugosa*) is a medium-sized shrub growing to 3 m tall. Twigs yellowish to reddish green, rough-textured (warty), streaked with purple. Relatively large white, pithy interior to stems. Leaves 7–15 cm long, 5–12 cm wide, veins arranged in

7–9 pairs curving toward the tip. Fruit light blue to greenish-white, to 6 mm. Inhabits sandy, rocky and gravelly slopes in southern, eastern Ontario, in the southwest and rarely in the north.

Silky dogwood (*C. obliqua*) is a spreading shrub to 3 m tall, branchlets hairy and grey, older stems with smooth, reddish to purple brown bark. Leaves more than twice as long (5–10 cm) as broad, smooth-edged, tapered tip. Clusters of small, cream-white flowers yielding round, blue fruits in clusters on long stalks. Widely distributed through sourthern Ontario in low, wet soils in marshes, thickets, open woodlands or near streams. Also called: silky cornel.

Red-osier dogwood (*C. sericea*)

Red-osier dogwood (*C. sericea*)

Huckleberries *Vaccinium* spp. and *Gaylussacia baccata*

Black huckleberry (*V. membranaceum*)

Huckleberries are delicious and well worth identifying and eating. These berries are generally sweet but can also be tart, strong-flavoured or seedy (blueberries, on the other hand, tend to be purely sweet and cranberries sharply tart). Huckleberry fruit are typically red, blackish and glossy, while blueberries are generally blue in colour with a whitish bloom on the fruit. Huckleberries can be used like domestic blueberries; they can be eaten fresh from the bush, added to fruit salads, cooked in pies and cobblers, made into jams and jellies or crushed and used in cold drinks. They are also delicious in pancakes, muffins, cakes and puddings.

Black huckleberries (*V. membranaceum* and *G. baccata*) were considered good for the liver and blood by some First Nations and were eaten as a ceremonial food to ensure health and prosperity for the coming season. The berries were traditionally eaten fresh, dried, made into cakes, or mixed with maple syrup as a drink. They were also combined with grease or oil to preserve

Black huckleberry (*V. membranaceum*)

them for winter use, gifts or as a trade item. The berries were mashed and formed into cakes or spread loosely on mats for drying. Later, they were reconstituted by boiling either alone or with other foods. Dried black huckleberry leaves and berries also make excellent teas.

Eating huckleberries can be sweet and wonderful but picking them can eat up a lot of time. Try using a (clean) comb to rake the berries into a basket, hat or other container to speed up the process. They can grow quite abundantly in some areas, so a little persistence can yield a good haul. Try to get there early, though—this berry is a favourite food of chipmunks, red and grey foxes, squirrels, skunks and many bird species such as blue, spruce and ruffed grouse as well as ptarmigan, bluebirds and thrushes. The plants themselves are a favourite browse for deer, bears and moose. Tough competition indeed!

EDIBILITY: unpalatable to highly edible

FRUIT: Berries (or drupes) range in colour from red to purple to black.

SEASON: Flowers April to June. Fruits ripen July to September.

DESCRIPTION: Deciduous or evergreen shrubs, 30 cm–2 m tall, with alternate, 2–5 cm-long leaves. Flowers various shades of pink, round to urn- or bell-shaped, 4–6 mm long, nodding on single, slender stalks. Fruits 6–10 mm across with a small "crown" at the bottom end of the berry, which is a residual left over from the flower that was pollinated to form the fruit. (This trait is found in all *Vaccinium* species.)

Black huckleberry (*V. membranaceum*) is a deciduous shrub to 2 m high with long-pointed elliptical leaves about 5 cm long and creamy-white bell-shaped flowers producing purplish black berries. Branches greenish when young but greyish brown with age and at most slightly angled. Grows in moist and acidic soils of forests and bogs. Also called: thinleaf huckleberry, black mountain huckleberry.

Black huckleberry (*V. membranaceum*)

Black huckleberry (*V. membranaceum*)

107

Black huckleberry (*G. baccata*) is a deciduous, low-growing, colonial shrub, 30 cm–1 m tall. Similar to blueberries and other huckleberries but with resin glands on the leaves, and fruits are berry-like drupes with only 10 seeds in each (compared with "true" berries of Vaccinium spp., with many seeds in each). Branchlets covered in fine hairs, brown. Older stems black to purplish grey, peeling. Leaves simple, elliptic to oblong to lance-shaped, alternate, entire, pinnately veined, and glandular above and

Tom's Huckleberry Pie

Makes 1 double-crust pie

The secret to a good, crispy pastry that is not tough and "dough-like" is to keep all your ingredients cool. Warm ingredients melt the small fat globules, causing them to mix too completely with the flour and resulting in chewy pastry. Leftover pastry trimmings make excellent little tartlets if rolled out again and put into the bottom of muffin tins, and filled with any extra huckleberry filling or jam out of the fridge.

Pastry
2 cups all-purpose flour
½ tsp salt
⅔ cup vegetable shortening
⅓ cup COLD milk

Filling
3 cups red or black huckleberries
¼ cup water or freshly squeezed orange juice
1 cup granulated white sugar
3 Tbsp cornstarch

For the pastry, sift the flour and salt together, then use 2 butter knives or a pastry cutter to cut the shortening into the flour mixture until the butter pieces are the size of small peas. *Avoid using your hands at this stage because their warmth will cause the butter to melt.* Gradually stir in the cold milk, then quickly shape the dough into 2 balls. Wrap them in plastic film, press into flat rounds and refrigerate immediately.

For the pie filling, mash the huckleberries with the water and put into a medium saucepan. Mix the cornstarch and sugar together and stir well into the COLD berries and water. *Do not heat the berries and water first because the cornstarch will cook prematurely and go all nasty and lumpy!* Bring mixture slowly to the boil, stirring constantly to avoid the cornstarch sticking or getting lumps. Simmer until noticeably thick, about 4 to 5 minutes, then take the saucepan off the burner.

Preheat oven to 350° F. Take the pastry out of the fridge, spread a thin layer of flour on a work surface, and roll the pastry until it is approximately ¼ inch thick. Place it into an 8-inch pie tin, cutting any extra off from around the edges. Then roll out the second half of your dough into a similar-sized round. Fill the pastry shell with the huckleberry mixture, then carefully place the second round on top. Gently push the edges of the top and bottom pastry crusts together (you may need to lightly wet one edge to get it to stay together), and prick a few holes with a fork in the top to allow steam to escape during cooking. Bake for approximately 50 minutes.

below. Flowers pink, 5-parted, about 6 mm wide, tubular, stalked, borne on a glandular raceme that is usually shorter than the leaves. Fruit reddish purple to black, 5 mm in diameter. Grows in dry, rocky or sandy soil in woods, forests and thickets, occasionally found in wetter areas. Also called: crackleberry.

Deerberry (*V. stamineum*) is a slender shrub to 1 m tall, with elliptical leaves about twice as long (to 9 cm) as wide, with blunt or pointed ends. Fruits yellowish green to blue, round, but unpalatable. A species at risk not found in other provinces, deerberry grows in dry, rocky woods and clearings almost exclusively in the Thousand Islands area of the St. Lawrence River. It should not be collected. Also called: squaw huckleberry.

COMPARE: *As suggested by their shared common name,* V. membranaceum *and* G. baccata *could easily be confused with each other or with wild blueberry species. To distinguish between black huckleberries, look for* G. baccata's *distinctive seedy fruits and golden-yellow resin glands on the leaves. Clearly, common names are not always consistent and can cause confusion. Take heart, though: both black huckleberries are edible and tasty!*

Black huckleberry (*G. baccata*)

Black huckleberry (*G. baccata*)

Blueberries *Vaccinium* spp.

Velvet-leaf blueberry (*V. myrtilloides*)

Blueberry fruit tend to be blue in colour, hence the common name of this group of plants. The fruit is generally sweet, rather than tart/sour (cranberries) or sweet/tangy (huckleberries). Wild blueberries are simply delicious! Rich in vitamin C and natural antioxidants, these fruit are both beautiful to look at and good for you to eat. The sweet, juicy berries can be eaten fresh from the bush, added to fruit salads, cooked in pies, tarts and cobblers, made into jams, syrups and jellies, or crushed and used to make juice, wine, tea and cold drinks. Blueberries also make a prime addition to pancakes, muffins, cakes and puddings as well as to savoury treats like chutneys and marinades.

These wild fruits were widely used by First Nations, either fresh, dried singly or mashed and formed into cakes. To make dried cakes, the berries were cooked to a mush to release the juice, spread into slabs and dried on a rack (made from wood, rocks, or plant materials) in the sun or near a fire. Often, the leftover juice was slowly poured onto the drying cakes to increase their flavour and sweetness. Some northern indigenous peoples stored boiled blueberries mixed with oil and later whipped this mixture with snow and fish oil to make a dessert.

Because blueberries grow close to the ground, they can be difficult and time-consuming to collect, so some people developed a method of combing them from the branches with a salmon backbone or wooden comb or rake. While this method is efficient, it results in baskets full of both berries and small hard-to-pick-out blueberry leaves. The savvy solution developed for this problem was to place a wooden board at a medium angle and slowly pour the berry/leaf mix from the top of the board: the berries (which are round) roll down the board to a basket waiting below, but the leaves (which are flat) stick to the board and stay put. After 2–3 rollings, the picker ends up with a basket of pure berries at significantly less effort than would have been required to pick the leaves out individually. The only drawback to this method is that the occasional green berry also gets picked, but these are easily removed by hand.

While most people only associate blueberries with a delicious fruit, there are many historical medicinal uses for other parts of this wide-ranging plant.

Blueberry roots were boiled to make medicinal teas that were taken to relieve diarrhea, gargled to soothe sore mouths and throats, or applied to slow-healing sores. Bruised roots and berries were steeped in gin, which was taken freely (as much as the stomach and head could tolerate!) to stimulate urination and relieve kidney stones and water retention. Blueberry leaf tea and dried blueberries have historically been used like cranberries to treat diarrhea and urinary tract infections. Blueberry leaf tea has also been used by people suffering from hypoglycemia and by some diabetics, to stabilize and reduce blood sugar levels, and to reduce the need for insulin. Blueberries contain anthocyanins, which are said to improve night vision. These compounds are most concentrated in

Highbush blueberry (*V. corymbosum*)

Dwarf blueberry (*V. caespitosum*)

Vaccinium spp.

Note that common names can be confusing, especially with the large variety of Vaccinium *species native to Ontario. Blueberries, cranberries and huckleberries are all closely related plants in this plant family. In North America there are approximately 35 different* Vaccinium *species, but hybridization is common in the genus so the true number of varieties is probably greater. As a general rule, species of* Vaccinium *with blue fruits are called blueberries or bilberries, and taller shrubs with fruits that aren't blue are called huckleberries. Shorter* Vaccinium *species with red berries and a distinctive tart flavour are commonly referred to as cranberries. However, as illustrated by black huckleberry, common names do not necessarily follow this botanical protocol. Another example is "high bush cranberry,"* Viburnum edule, *which is in the honeysuckle family and is not a "true" cranberry at all, despite its red colour and tart flavour. Rest assured, though, none of these species is poisonous and they are all delicious to eat!*

the dried fruit, preserves, jams and jellies. Their effect, however, is said to wear off after 5–6 hours. Anthocyanins may also reduce leakage in small blood vessels (capillaries), and blueberries have been suggested as a safe and effective treatment for water retention during pregnancy, for hemorrhoids, varicose veins and similar problems. They have also been recommended to reduce inflammation from acne and other skin problems and to prevent cataracts. The leaf or root tea of low bilberry is reported to flush pinworms from the body.

The leaves of blueberry were sometimes dried and smoked, and the berries have been used to dye clothing a navy blue colour.

EDIBILITY: highly edible

FRUIT: Berries round, 5–8 mm wide, bluish-coloured, growing in clusters, usually with a greyish bloom.

Velvet-leaf blueberry (*V. myrtilloides*)

Bog blueberry (*V. uliginosum*)

Dwarf blueberry (*V. caespitosum*)

Oval-leaved blueberry (*V. ovalifolium*)

Season: Flowers May to July. Fruits ripen July to October.

Description: Low, often matted shrubs with thin, oval leaves 1–7.5 cm long. Flowers whitish to pink, nodding, urn-shaped, 4–6 mm long.

Bog blueberry (*V. uliginosum*) is a low, spreading, deciduous, perennial shrub, 10–60 cm tall (but can grow as short as 2.5 cm in areas of heavy snow where the shrub is crushed flat each winter). Branches are rounded, brownish. Leaves 3 cm long, alternate, fuzzy to smooth, elliptical to oval, dull whitish green, toothless and broadest toward the tip, which may be rounded, pointed or indented. Flowers solitary or paired, 4–5 lobed, white or pinkish, urn-shaped, up to 6 mm long. Berries dark blue to blackish with a whitish bloom, to 9 mm in size. Inhabits low elevation bogs, boggy forests and rocky or sandy lake and river shores in northern and western Ontario. Also called: bog bilberry, western huckleberry • syn. *V. occidentale*.

Dwarf blueberry (*V. caespitosum*) is a low, usually matted shrub, 10–30 cm tall, with rounded, yellowish to reddish branches and finely toothed, light green leaves to 2.5 cm long, widest above the middle. Its 5-lobed flowers produce blue-black berries (6 mm) singly in leaf axils, from August to September. Grows in rocky and sandy clearings and boreal thickets of northern and western Ontario.

High bush blueberry (*V. corymbosum*) is the tallest blueberry, a multistemmed shrub growing 2–4 m high with older stems red to black and branchlets green to brown. Leaves to 7.5 cm long, about three times the width. Flowers 5-toothed and white in late spring to produce blue or blue-black fruits from August to October. Grows in woodland and the edges of swamps or ponds in southern and eastern Ontario.

113

Low-bush blueberry (*V. angustifolium*) is a low shrub 30–60 cm tall that spreads into large patches. Leaves alternate, elliptical, to 45 mm long and 15 mm wide, generally hairless on both sides. White, bell-shaped flowers 5 mm long, from May to June, and small, dark-blue or black berries, up to 1.2 cm across. Grows in abundance in dry, open barrens, peats and rock and is the main source of wild harvested blueberries in eastern Canada (including Ontario). Also called: low sweet blueberry.

Oval-leaved blueberry (*V. ovalifolium*) is a shrub, to 2 m tall but often smaller, with hairless, angled branches. Leaves are entire or only sparsely toothed and more broadly oval than other blueberry species. Produces dark blue or purplish berries with a whitish bloom from early July to September. Grows at low elevations in mixedwood

Low-bush blueberry (*V. angustifolium*)

forests and along the shoreline of eastern Lake Superior.

Velvet-leaf blueberry (*V. myrtilloides*) is a low shrub very similar to low bush blueberry except with densely hairy branches, especially when young, and downy leaves (velvety). Produces blue or dark bluish-black fruits from August to October. Grows throughout the province in dry to moist forests, clearings and bogs.

WARNING: *Blueberry leaves contain moderately high concentrations of tannins, so they should not be used continually for extended periods of time.*

Oval-leaved blueberry (*V. ovalifolium*)

Bog blueberry (*V. uliginosum*)

Dwarf blueberry (*V. caespitosum*)

Blueberry Cobbler

1 cup flour · 2 Tbsp sugar · 1½ tsp baking powder
¼ tsp salt · 1 tsp grated lemon zest · ¼ cup butter
1 beaten egg · ¼ cup milk · 2 Tbsp corn starch
½ cup sugar · 4 cups fresh blueberries (or huckleberries)

Preheat oven to 425° F. Sift together all the dry ingredients. Mix wet ingredients together, then pour slowly into the dry mix, stirring until just moistened.

Mix cornstarch and sugar together, and toss with the fruit. Pour this mixture into the bottom of an 8 x 10-inch glass or ceramic baking dish (avoid metal dishes because the acid in the fruit might turn it rusty and impart a nasty flavour to the cobbler). Drop the topping in spoonfuls on top of the fruit, covering the surface as evenly as possible (some exposed areas of fruit are fine). Bake, uncovered, for 25 minutes or until light brown.

Fruit Popsicles

Makes 8 to 12 popsicles

Easy and a popular treat for adults and kids at any time of the year!

4 cups wild berries · 1 cup plain yogurt or light cream
1 cup white sugar · 1 cup orange or other fruit juice (optional)

Blend all ingredients together, pour into the compartments of a popsicle maker and place in freezer until frozen.

Cranberries *Vaccinium* spp.

Lingonberry (*V. vitis-idaea*)

Like many other species of wild berries with domesticated counterparts, wild cranberries are small but packed with a flavour that seems disproportionate to their size. The tartness of cranberries gives them an enviable versatility for sweet, sour and savoury dishes. Who could imagine Thanksgiving dinner or many juices, desserts or baking without them? These berries, which also have a long history of medicinal use in treating kidney and urinary ailments, are touted today for their strong antioxidant properties. (Note: highbush cranberries, while tart and cranberry-tasting, are in the honeysuckle family and are therefore treated in a separate account.)

Cranberries can be very tart, but they make a refreshing trail nibble. They make delicious jams and jellies, and they can be crushed or chopped to make tea, juice or wine. A refreshing drink is easily made by simmering the berries (crush them first to allow the juice out more easily) with sugar and water or, more traditionally, by mixing cranberries with maple sugar and cider. Cranberry sauce is still a favourite with meat or fowl, and the berries add a pleasant zing to fruit salads, pies and mixed fruit cobblers. Cranberries are also a delicious addition to pancakes, muffins, breads, cakes and puddings.

Firm, washed berries keep for several months when stored in a cool place. They can also be frozen, dried or canned. First Nations sometimes dried cranberries for use in pemmican, soups, sauces and stews. Some tribes stored boiled cranberries mixed with oil and later whipped this mixture with snow and fish oil to make a dessert.

Freezing makes cranberries sweeter, so they were traditionally collected after the first heavy frost and can be foraged when snows melt in the spring—if the wildlife have left any behind! Because they remain on the shrubs all year, cranberries can be a valuable survival food rich in vitamin C and antioxidants. These low-growing berries are difficult to collect, so some people combed them from branches with a salmon backbone or wooden comb.

Cranberry juice has long been used to treat bladder infections. Research shows that these berries contain arbutin and procyanidins (a specific type of tannin), which prevents some bacteria from adhering to the walls of the bladder and urinary tract and causing an infection. Cranberry juice also increases the acidity of the urine, thereby inhibiting bacterial activity,

which can relieve infections. Commercial cranberry juice or cocktail blends are not appropriate for this treatment, however, as the juice is highly processed, often diluted with other juices and is highly sweetened (sugars will feed the problem bacteria and can make the existing condition worse). Increased acidity can also lessen the urinary odour of people suffering from incontinence. The tannins in cranberries have anti-clotting properties and can reduce the amount of dental plaque-causing bacteria in the mouth, thus are helpful against gingivitis. Research has shown that cranberries contain antioxidant polyphenols that may be beneficial in maintaining cardiovascular health and immune function, and in preventing diabetes and tumour formation. Although some of these compounds have proved extraordinarily powerful in killing certain types of human cancer cells in the laboratory, their effectiveness when ingested is unknown. There is also evidence that cranberry juice may be

Large cranberry (*V. macrocarpon*)

effective against the formation of kidney stones.

Cranberries were traditionally prescribed to relieve nausea, to ease cramps in childbirth and to quiet hysteria and convulsions. Crushed cranberries were used as poultices on wounds, including poison-arrow wounds. The red pulp (left after the berries have been crushed to make juice) can be used to make a red dye. A related species, lingonberry, is popular in Sweden as a digestive aid and has been used in jams, jellies, pies and other baking, juice, wine liqueur, herbal tea, and as "cranberry sauce."

EDIBILITY: highly edible

FRUIT: Berries bright red (sometimes purplish) when ripe, 6–10 mm wide.

SEASON: Flowers June to July. Fruits ripen late July to September and may persist on plants throughout winter.

DESCRIPTION: Dwarf, low-spreading shrubs or vines, usually less than

Bog cranberry (*V. oxycoccos*)

20 cm tall, often trailing, with leathery leaves less than 15 mm long. Small, nodding, 4-parted, pinkish flowers produce sour, bright red (sometimes purplish) cranberries.

Bog cranberry (*V. oxycoccos*) is a vine with slender stems, creeping, often

Large cranberry (*V. macrocarpon*)

Bog cranberry (*V. oxycoccos*)

Bog cranberry (*V. oxycoccos*)

Cranberry Chicken

Serves 5

3 lbs chicken • ¼ cup flour • ½ tsp salt
¼ cup cooking oil • 1½ cups fresh cranberries
½ cup sugar • 1 Tbsp grated orange zest
½ cup fresh orange juice • ¼ tsp ground ginger

Cut chicken into serving-sized pieces, and coat with flour and salt. Heat oil in a cast-iron skillet. Add chicken pieces and brown on both sides, being careful to cook the chicken fully. Combine remaining ingredients in a saucepan, bring to the boil and pour it over the chicken in the skillet. Cover skillet, reduce heat and simmer 30 to 40 minutes until chicken is tender.

rooting at nodes and reaching 1 m or more in length with small (less than 1 cm long), pointed, glossy leaves. Flowers are distinctive, 4 petals separated almost to the base, petals curve strongly backward (like little shooting stars), appears terminal on stems. Fruit a deep red, globose cranberry, 5–12 mm wide, from July to August. Bog cranberry predictably inhabits bogs (and less predictably tundra). Also called: small cranberry • syn. *Oxycoccus oxycoccos, O. quadripetalus, O. microcarpus.*

Large cranberry (*V. macrocarpon*) has separated petals almost to the base that spread or curve backward in clusters arising from the leaf axils. Stems slender, wiry and intertwining. Leaves 5–15 mm long. Fruit a dark red, globose berry, 10–20 mm wide, appear from July to August and remain through winter. Grows at low elevations in open bogs, swamps and the shores of lakes, ponds and streams from eastern Ontario south to Windsor and west to Lake Superior. Also called: syn. *Oxycoccus macrocarpus.*

Lingonberry (*V. vitis-idaea*) is a low-spreading shrub with rounded or slightly angled branches. Leaves evergreen, 6–15 mm long, blunt, leathery, with dark dots (hairs) on a pale lower surface. Flowers 4-petalled, fused into small urn-shaped nodding blooms, pinkish in colour, 1 to several in terminal clusters. Fruit bright red cranberries, 6–10 mm wide, from July to August. Grows in rocky or sandy clearings and in acidic soils of peat bogs, dry woods and conifer forests. Also called: mountain cranberry, low-bush cranberry, rock cranberry, cowberry, partridge berry.

Lingonberry (*V. vitis-idaea*)

119

False Wintergreens   *Gaultheria* spp.

Hairy false-wintergreen (*G. hispidula*)

The small, sweet berries of this species are delicious and can be eaten fresh, served with cream and sugar or cooked in sauces. Their flavour improves after freezing, so they are at their best in winter following the first frost (even from under the snow if you are persistent!), or in spring when they are plump and juicy. The young leaves can be an interesting trail nibble or added to salads as well as used to make a strong, aromatic tea that is said to make a good digestive tonic. The wintergreen flavour can be drawn out if the bright red leaves are first fermented.

The berries were historically mixed with teas and were used to add fragrance and flavour to liqueurs. Occasionally, large quantities were picked and dried like raisins for winter use. During the American Revolutionary War, wintergreen tea was a substitute for black tea (*Camellia sinensis*). The berries were traditionally soaked in brandy, and the resulting extract was taken to stimulate appetite, as a substitute for bitters.

All false wintergreens contain methyl salicylate, a close relative of aspirin that has been used to relieve aches and pains. These plants were widely used in the treatment of painful, inflamed joints resulting from rheumatism and arthritis (see Warning). Studies suggest that oil of wintergreen is an effective painkiller and also has numerous commercial applications. It provides fragrance to various products such as toothpastes, chewing gum and candy and is used to mask the odours of some organophosphate pesticides. It is a flavouring agent (at no more than 0.04 percent) and an ingredient in deep-heating creams. The oil is also a source of triboluminescence, a phenomenon in which a substance produces light when rubbed, scratched or crushed. The oil, mixed with sugar and dried, builds up an electrical charge that releases sparks when ground, producing the Wint-O-Green Lifesavers optical phenomenon. To observe this, look in a mirror in a dark room and chew the candy with your mouth open!

EDIBILITY: highly edible

FRUIT: Fruit mealy to pulpy, fleshy, berry-like capsules with a mild, wintergreen flavour.

SEASON: Flowers May to June. Fruits ripen July into September.

DESCRIPTION: Delicate, creeping evergreen shrublets. Leaves leathery, small, alternate, oval to elliptical. Flowers white to greenish or pinkish.

Hairy false-wintergreen (*G. hispidula*) has tiny, stiff, flat-lying, brown hairs on its stems and lower leaf surfaces. Leaves very small (4–10 mm long) and numerous. Flowers tiny (2 mm wide), 4-lobed and bell-shaped. Berries white, small (generally less than 5 mm in diameter), on a very short stalk, persisting through fall. Grows in cold, wet bogs and coniferous forests. Widely distributed and associated with acidic soils, often grows in mosses under conifers or on rotting logs, along the edges of swamps or bogs. Also called: snowberry, creeping snowberry.

Wintergreen (*G. procumbens*) has thick, shiny, oval leaves 2.5–5 cm long, with small, oval flowers and red berries that dangle beneath the leaves and often stay on the plant through winter.

WARNING: *Oil-of-wintergreen (most concentrated in the berries and young leaves) contains methyl salicylate, a drug that has caused accidental poisonings. It should never be taken internally, except in very small amounts. Avoid applying the oil when you are hot, because dangerous amounts could be absorbed through the open pores of your skin. It is known to cause skin reactions and severe (anaphylactic) allergic reactions. People who are allergic to aspirin should not consume or use false wintergreen or its relatives.*

Grows in poor or sandy soils under evergreens or in oak woods across the province south of 50° N. Also called: Eastern teaberry, checkerberry.

Wintergreen (*G. procumbens*)

Hairy false-wintergreen
(*G. hispidula*)

Bearberries *Arctostaphylos* spp.

Common bearberry (*A. uva-ursi*)

Bearberries are rather mealy and tasteless, but they are often abundant and remain on plants all year, so they can provide an important survival food. Many First Nations traditionally ate these berries. To reduce the dryness, bearberries were often cooked with oil, bear fat or fish eggs, or they were added to soups or stews. Sometimes, boiled berries were preserved in oil and served whipped with snow during winter. Boiled bearberries, sweetened with syrup or sugar and served with cream, reportedly make a tasty dessert. They can also be used in jams, jellies, cobblers and pies, or dried, ground and cooked into a mush. Apparently, if the berries are fried in grease over a slow fire they eventually pop, rather like popcorn. Scalded mashed berries, soaked in water for an hour or so, produce a spicy, cider-like drink, which can be sweetened and fermented to make wine.

Although fairly insipid, juicy alpine bearberries are probably among the most palatable fruits in the genus, but because they grow at high elevations and northern latitudes, they have been the least used. Hikers sometimes chew bearberries and the leaves to stimulate saliva flow and to relieve thirst.

All parts of the bearberry plant were widely used medicinally by Native groups across Canada and the U.S. for purposes varying from pain relief to treating kidney problems or pimples. Today, the leaves are commonly used and sold as tea to treat urinary infections.

EDIBILITY: edible, not palatable

FRUIT: Small 5–10 mm fruits, bright red to purplish black.

SEASON: Flowers May to July. Fruits ripen August to September.

DESCRIPTION: Evergreen or deciduous prostrate shrubs with simple, alternate leaves and clusters of nodding, white or pinkish, urn-shaped flowers and juicy to mealy, berry-like drupes containing 5 small nutlets.

Alpine bearberry (*A. alpina*) is a trailing, deciduous shrub, to 15 cm tall. Leaves thin, veiny, oval, 1–5 cm long, with hairy margins (at the base) that often turn red in autumn, the previous year's dead leaves usually evident. Flowers small (4–6 mm long), appear from June to July and produce mealy fruits that are purplish black, 5–10 mm in diameter, by late summer. Grows in moderately well-drained, rocky, gravelly and sandy soils on tundra, slopes and ridges in northern parts of Ontario. Also called: black alpine bearberry, whortleberry • syn. *Arctuous alpina*.

Common bearberry (*A. uva-ursi*) is a trailing, evergreen shrub to 15 cm tall. Leaves leathery, evergreen, spoon-shaped, 1–3 cm long. Small (4–6 mm long) flowers appear from May to June and produce bright red, 5–10 mm diameter, mealy fruits by late summer. Grows in well-drained, often gravelly or sandy soils in open woods and rocky, exposed sites at all elevations throughout Ontario. Also called: kinnikinnick • syn. *Arctuous rubra*.

Red bearberry (*A. alpina* var. *rubra*) is similar to alpine bearberry, but it has longer leaves (to 9 cm long) with hairless margins, the leaves of the previous year not persistent. Berries bright red. Grows in the same habitats as alpine bearberry and over the same range.

Alpine bearberry (*A. alpina*)

Common bearberry (*A. uva-ursi*)

Red bearberry (*A. alpina* var. *rubra*)

Black Crowberry *Empetrum nigrum*

Also called: moss berry, curlew berry

Black crowberry (*E. nigrum*)

Next to cranberries and blueberries, crowberries are one of the most abundant, edible wild fruits found in northern Canada and were a vital addition to the diet of northern First Nations. Because crowberries are almost devoid of natural acids, they can taste a little bland and were often mixed with blueberries or lard or oil and in more modern times with sugar and lemon. Their taste does seem to vary greatly with habitat—the flavour of the berries has been described in a range from bland to tasting like turpentine and even most delicious. Their taste improves after freezing or cooking, however, and their sweet flavour peaks after a frost.

The fruit is high in vitamin C, about twice that of blueberries, and is also rich in antioxidant anthocyanins (the pigment that gives the berries their black colour). Their high water content was a blessing to hunters seeking to quench their thirst in the waterless high country. As the berries have a firm, impermeable skin and are not prone to becoming soggy, they are ideal for making muffins, pancakes, pies, jellies (with added pectin), preserves and the like. For a fine dessert, cook the berries with a little lemon juice and serve them with cream and sugar.

Black crowberry (*E. nigrum*)

Black crowberry (*E. nigrum*)

Crowberries are usually collected in autumn, but because they often persist on the plant over winter, they can be picked (snow depth permitting) through to spring if the wildlife doesn't get them first. They are small fruit, though, and it can take up to 1 hour to pick 2 cups of berries! Consuming too many crowberries may cause constipation, so the berries have historically been prescribed for diarrhea. The berries make a reasonable black dye.

EDIBILITY: highly edible

FRUIT: Black, shiny, berry-like drupe to 9 mm. Contains large, inedible seeds.

SEASON: Flowers May to August. Fruits ripen July to November.

DESCRIPTION: Evergreen dwarf or low shrub, 5–10 cm tall, prostrate and mat-forming, to more than 1 m across. Leaves dark to yellowish-green to wine-coloured, 2–6 mm long, alternate but growing so closely together as to appear whorled, needle-like. Flowers inconspicuous, 1–3, pink, in leaf axils, 3 petals and sepals, petals 3 mm long, with male and female flowers separate but on the same plant. Fruit a juicy, black, berry-like drupe containing 2–9 seeds, sometimes overwintering. Grows prolifically in bogs, moist shady forests, acidic, rocky or gravelly soil on slopes, ridges and seashores in tundra and spruce forests of western and northern Ontario.

Riverbank Grape   *Vitis riparia*

Also called: frost grape

Riverbank grape (*V. riparia*)

The fruit of this wild grape is small and tart, but juicy and flavourful. Many First Nations valued the berries (eaten fresh or preserved for winter) as a food source. These grapes are said to be best when harvested after the first frost, which makes them taste sweeter, but they are often consumed rapidly by birds, so gatherers need to be quick! In more recent times, the berries have been made into jelly and wine. The wine of our native grapes reportedly has a musky, "foxy" flavour. Wine and grapeseed extract contain the compound resveratrol, which has beneficial cardiovascular and antidiabetic properties, as well as strong antioxidants.

The winter vines, when twisted together, make a solid and decorative base for Christmas wreaths. The leaves can be pickled to use in making dolmades, a Greek delicacy of rice and meat wrapped in vine leaves. Did your grandmother insist on stuffing a grape

COMPARE: *Two other native grape species are uncommonly found in the Niagara region. Fox grape (*V. labrusca*), from which the concord grape is derived, has bigger fruits (to 2 cm) and leaves with rusty-coloured hairs on the undersides. Summer grape (*V. aestivalis*) grows taller than both riverbank and fox grape (to 10 m) and its leaves are whitish to silvery beneath.*

leaf in each jar of pickles she made? The leaves contain varying amounts of a natural inhibitor that reduces the effects of a softening enzyme present on mouldy cucumber blossoms. Adding a grape leaf to each jar of homemade pickles is a traditional practice that really does result in pickles that are less likely to go soft.

Riverbank grape is a key parent species in breeding modern grape varieties for fruit and wine that are disease resistant and cold tolerant.

FRUIT: Fruits a tight cluster of black, spherical berries with a waxy coating, giving them a bluish cast; small, 10–12 mm across, containing 2–6 small, oval seeds each.

EDIBILITY: highly edible

SEASON: Flowers May to June. Fruits ripen August to September.

DESCRIPTION: A woody, deciduous vine either climbing by means of tendrils to 5 m high or trailing on the ground. Leaves alternate, 3–5 lobed, 7–20 cm long, hairless or slightly hairy below, coarsely toothed edges. Flowers greenish, inconspicuous; compact in pyramidal clusters. Inhabits thickets, open woods, river banks and rocky or sandy soils. Common in southern Ontario.

Riverbank grape (*V. riparia*)

WARNING: *Beware not to confuse this species with another vine that is related but only superficially similar in appearance: moonseed (*Menispermum canadense*). The dark blue, berry-like fruits of moonseed are highly poisonous and may be mistaken for edible grapes, particularly by children. The leaves of moonseed are smooth rather than toothed, and each fruit contains one crescent-shaped seed rather than the many seeds found in wild grapes.*

Riverbank Grape Jelly

Makes about 4 x 1 cup jars

4 cups ripe riverbank grapes, de-stemmed
1 cup water • unsweetened apple juice
1 x 2 oz package pectin • 5 cups sugar

Clean and crush grapes. Place water and crushed grapes into a heavy-bottomed saucepan and cook slowly for about 20 minutes until fruit is softened and letting its juices go. Strain mixture through a jellybag or cheesecloth. *Do not squeeze the cloth or it will cause sediments to run into the juice, resulting in a cloudy jelly.* If the resulting liquid does not measure 4 cups, add some apple juice until 4 cups liquid is reached. Add liquid back to clean saucepan. Mix in pectin and bring the mixture to a full boil. Add sugar and continue stirring, scraping the bottom of the pot so that the jelly doesn't burn. Boil hard for 3 minutes, then remove from heat and pour into hot sterilized jars. Wipe jar edges clean and seal with hot sterilized lids (waterbathing them for 5 minutes will also seal jars).

Elderberries *Sambucus* spp.

Red elderberry (*S. racemosa*)

Raw elderberries are generally considered inedible and cooked berries edible (see Warning), but some tribes are said to have eaten large quantities fresh from the bush. Cooking or drying the berries destroys the rank-smelling, toxic compounds. Most elderberries were consumed after steaming or boiling, or were dried for winter use. Sometimes clusters of fruit were spread on beds of pine needles in late autumn and covered with more needles and eventually with an insulating layer of snow. These caches were easily located in the winter months by the bluish-pink stain they left in the snow! Only small amounts were eaten at a time, though, just enough to get a taste.

Red elderberry (*S. racemosa*)

Today, elderberries are used in jams, jellies, syrups, preserves, pies and wine. Because these fruits contain no pectin, they are often mixed with tart, pectin-rich fruits such as crab apples to ensure that the jam or jelly sets. Elderberries are also used to make teas and to flavour some wines (e.g., Liebfraumilch) and liqueurs (e.g., Sambuca). A delicious, refreshing fizzy drink called elderflower presse or cordial can be made from the flowers. The flowers can be used to make tea or wine, and in some areas, flower clusters were popular dipped in batter and fried as fritters or stripped from their relatively bitter stalks and mixed into pancake batter.

Elderberries are rich in vitamin A, vitamin C, calcium, potassium and iron. They have also been shown to contain antiviral compounds that could be useful in treating influenza. The berries can be used to produce a brilliant crimson or violet dye. Elderberry wine, elderberries soaked in buttermilk and elderflower water have all been used in cosmetic washes and skin creams. The scientific name *Sambucus* derives from the Greek instrument sambuke, in reference to the hol-

Common elderberry (*S. nigra*)

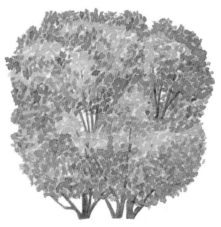

Common elderberry (*S. nigra*)

low pithy stems of this plant, which have been used in many different cultures to make musical instruments.

EDIBILITY: edible, edible with caution (toxic)

FRUIT: Fruits juicy, berry-like drupes, 4–6 mm across, in dense, showy clusters.

SEASON: Flowers April to July. Fruits ripen July to September.

DESCRIPTION: Unpleasant-smelling, 1–4 m tall, deciduous shrubs with

pithy, opposite branches often sprouting from the base. Leaves pinnately divided into 5–9 sharply toothed leaflets about 5–15 cm long. Flowers white, 4–6 mm wide, forming crowded, branched clusters.

Common elderberry (*S. nigra*) is a shrub similar to red elderberry but with a white pith. It bears large corymbs (20–30 cm across) of white flowers and dark purple to black berries. Grows in rich, moist soils along streams and rivers throughout southern Ontario. Also called: American elder.

Red elderberry (*S. racemosa*) has pyramid-shaped flower clusters and shiny red fruits that are considered the tastiest of the genus. Grows to 4 m tall, soft or slightly woody stems with an orange or red-brown pith. Inhabits open woods, forest edges and roadsides or ravines across southern Ontario to west of Lake Superior, infrequently farther north. Also called: red-berried elder • *S. pubens, S. melanocarpa.*

Red elderberry (*S. racemosa*)

WARNING: *All parts of this plant, except for the fruit and flowers, are considered toxic. The stems, bark, leaves and roots contain poisonous cyanide-producing glycosides (especially when fresh) that can cause nausea, vomiting and diarrhea, but the ripe fruits and flowers are edible. The seeds, however, contain toxins that are most concentrated in red-fruited species. Many sources classify red-fruited elderberries as poisonous and black- or blue-fruited species as edible.*

Red elderberry (*S. racemosa*)

Common elderberry (*S. nigra*)

Red elderberry (*S. racemosa*)

Bush-cranberries *Viburnum* spp.

High bush cranberry (*V. edule*)

These berries are best picked in autumn, after they have been softened and sweetened by freezing. Some people compare their fragrance to that of dirty socks, but the flavour is good (perhaps a Stilton of the berry world?). The addition of lemon or orange peel is said to eliminate the odour. Raw bush-cranberries can be very sour and acidic (much like true cranberries), but many Native peoples such as the Abenake and Algonquin ate them, chewing the fruit, swallowing the juice and spitting out the tough skins and seeds. They were also eaten with bear grease, or, in an early year, they could be mixed with sweeter berries such as strawberries. Bush-cranberries are an excellent winter-survival food because they remain on branches all winter and are much sweeter after freezing. Some tribes ate the boiled berries mixed with oil and occasionally this mixture was whipped with fresh snow to make a frothy dessert. Today, bush-cranberries are usually boiled, strained (to remove the seeds and skins) and used in jams and jellies. While these preserves usually require additional pectin (especially after the berries have

Marsh cranberry (*V. trilobum*)

European highbush cranberry (*V. opulus*)

been frozen), there are reports that imperfectly ripe berries (not yet red) jell without added pectin. Bush-cranberry juice can be used to make a refreshing cold drink or fermented to make wine, and the fresh or dried berries can be steeped in hot water to make tea. Unfortunately, their large stones and tough skins limit their use in muffins, pancakes and pies unless the fruit is strained first. The berries produce a lovely reddish-pink dye and the acidic juice can be used as a mordant (required to set dyes and make the colour permanent).

The bark is said to have a sedative effect and to relieve muscle spasms, so it has been widely used to treat menstrual pains, stomach cramps and sore muscles. Bush-cranberry bark has also been used to relieve asthma, hysteria and convulsions, and it was sometimes given to women threatened with miscarriage (to stop contractions). There is some controversy over bush-cran-berry's ability to relieve cramps, but some sources report that pharmacological research supports this use. Commercial "crampbark" was sometimes really the bark of mountain maple (*Acer spicatum*) because of mistakes made by collectors. A few people smoked bush-cranberry bark.

American bush-cranberry (European highbush cranberry) makes a superb garden ornamental that is drought tolerant and provides not only pretty and scented spring flowers but also a

133

showy fall foliar display and important winter wildlife food (if the humans don't get there first!).

EDIBILITY: highly edible

FRUIT: Fruits juicy, strong-smelling, red to orange or black, berry-like drupes, 1–1.5 cm long, with a single flat stone.

SEASON: Flowers April to July. Fruits ripen July to October.

DESCRIPTION: Deciduous, unarmed shrubs with opposite, entire or 3-lobed leaves up to 20 cm long. Flowers white, small, 5-petalled, forming flat-topped clusters.

Downy arrowwood (*V. rafinesquianum*) grows to 2 m tall but is usually shorter. Leaves 4–8 cm long, 3–5 cm wide, prominently toothed with stipules, dark green above and densely hairy below. Fruit purplish-black, to 1 cm in diameter. Grows in the understorey of the boreal forest in northwestern Ontario in thickets and woods or on slopes and river banks of southern Ontario.

WARNING: *Some sources classify raw bush cranberries as poisonous, while others report that they were commonly eaten raw by native peoples. A few berries may be harmless, but large quantities can cause vomiting and cramps especially if they are not fully ripe, so it is probably best to cook the fruit before eating. Despite the common name of "cranberry," these species are not botanically related to the sour red berries we traditionally enjoy with a Thanksgiving feast.*

European highbush cranberry (*V. opulus*) is a large shrub, 1–4 m tall, with a wide-spreading habit. Leaves maple leaf–like with three relatively deep-cut lobes. Flower clusters resemble a lacecap hydrangea with a showy outer ring of large (12–25 mm wide), white, sterile flowers surrounding a central growth of tiny petal-less blooms. Bright orange to deep-red fruits in clusters, each up to 1 cm in diameter. Grows in moist soils, hedges, scrub areas and forest edges. Native to Eurasia but now occurs, escaped, in Ontario. Also called: American bush-cranberry, pembina • syn. *V. trilobum* var. *americanum*.

High bush cranberry (*V. edule*) is a scraggly-looking shrub reaching 3 m in height. Bark smooth, grey with a reddish tinge. Leaves opposite, 6–12 cm wide, sharply toothed and hairy underneath, often with three shallow lobes evident toward the leaf tip. Flowers small, relatively inconspicuous clusters growing beneath leaf pairs. This plant's distinctive, musty smell may announce its presence before it is actually seen. Fruits initially yellow maturing to orange or red. Grows in shady foothills, damp woods, thickets, streambanks and lakeshores of northern and western Ontario. Also called: mooseberry, squashberry, low bush cranberry.

Hobbleberry (*V. alnifolium*) is a 0.9–3 m tall shrub. Leaves heart-shaped, 7.5–20 cm long, with prominent veins, finely saw-toothed margins and star-like, rusty hairs beneath. Fragrant, flat-topped clusters (5–15 cm wide) of small, white flowers, the outer flowers larger (2.5 cm wide) than the

High bush cranberry (*V. edule*)

European highbush cranberry (*V. opulus*)

High bush cranberry (*V. edule*)

European highbush cranberry (*V. opulus*)

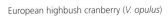

inner ones. Grows in moist woods and shady ravines south of Hamilton to Georgian Bay and Québec. Also called: hobble bush.

Mapleleaf viburnum (*V. acerifolium*) is a 0.9–1.8 m tall shrub with maple-like (but 3-lobed) leaves, downy beneath. Flowers white, 6 mm wide in flat-topped clusters 5–7.5 cm wide. Found in southern and eastern Ontario in dry or moist open thickets, hardwood forests and ravine slopes.

Marsh cranberry (*V. trilobum*) is a 3–4.5 m tall shrub with large, showy, white, sterile outer flowers in each cluster, and bright red, translucent berries in late summer and autumn. Leaves 5–11 cm long, sharply 3-lobed, each with pointed tips. Closely resembles European high bush cranberry but found more frequently in the wild. Grows in low, moist sites and along streambanks. Also called: high bush cranberry (yet again!)

Nannyberry (*V. lentago*) is a shrub, to 9 m tall, with long, tapering leaf tips and winged petioles. The wood has an unpleasant, goat-like smell. Leaves finely toothed, to 10 cm long and 7 cm wide with pointed tips. Fruits hang in open clusters, raisin-like, dull blue-black when ripe. Grows along woodland edges, streams and rocky hillsides or swamps and bogs, mostly in southern Ontario but also west of Lake Superior. Also called: sheepberry.

Northern wild raisin (*V. cassinoides*) is an erect, deciduous shrub spreading from rhizomes to form thickets, stiffly branched, top-spreading, to 5 m tall. Mature twigs are ridged and purplish in colour. Leaves 5–10 cm long,

Nannyberry (*V. lentago*)

Nannyberry (*V. lentago*)

2.5–5 cm wide, upper surface green with a pale vein, brown-hairy below, opposite, blunt-tipped, oval to oblong, margins toothless or wavy-toothed, on grooved stalks. Fruit 6–9 mm in diameter, turning from green or whitish-yellow to pinkish then bright blue before finally becoming blue-black with a powdery coating when ripe. Inhabits wet conifer swamps and upland pine and mixedwood sites in moist to dry, clay to sandy soils. Also called: withered, withe rod • syn. *Nudum cassinoides*.

European highbush cranberry (*V. opulus*)

European highbush cranberry (*V. opulus*)

Nannyberry (*V. lentago*)

Wayfaring Tree *Viburnum lantana*

Wayfaring tree (*Viburnum lantana*)

This introduced European species is appreciated as an ornamental tree with showy fall foliage, bright berries, pretty leaves and flowers and good winter wildlife value. It was originally limited to urban plantings (popular because of its foliage, drought tolerance and insect and disease resistance) but has naturalized in southern Ontario. The fruits are very astringent-tasting and are inedible raw. They do, however, make a reasonable jelly when cooked and sweetened, either alone or combined with other fruits. Mature berries were historically used to make an ink.

EDIBILITY: edible

FRUIT: Berry-like drupes, ripening red to black, 8–10 mm long, single stone, fleshy.

SEASON: Flowers May to June. Fruits ripen August to September.

DESCRIPTION: Large deciduous shrub or small tree to 4 m tall, similar to hobbleberry but more erect with a dense, rounded crown. Twigs and branches wide-spreading, yellowish-grey, hairy. Buds greyish, lack scales,

Wayfaring tree (*Viburnum lantana*)

2 pairs of young leaves, covered in yellowish-white hairs. Leaves longer than wide, dark dull green above, leathery, prominently veined, opposite, turning dark red in fall. Flowers creamy white, in clusters 5–10 cm wide, at branch tips.

Wayfaring tree (*Viburnum lantana*)

Wayfaring tree (*Viburnum lantana*)

Wayfaring tree (*Viburnum lantana*)

Soapberry *Shepherdia canadensis*

Also called: russet buffaloberry

Soapberry (*S. canadensis*)

Soapberries were an important fruit for the First Nations within the plant's range, either for home use or as a trade item. The often-abundant berries were eaten fresh, though sometimes bitter, or they were boiled, formed into cakes and dried over a small fire for future use. Because their juice is rich in saponin, soapberries become foamy when beaten. The ripe fruit were mixed about 4:1 with water and whipped like egg whites to make a foamy dessert called "Indian ice cream." The resulting foam is truly unexpected and remarkable, having a beautiful white to pale pink colour and a smooth, shiny consistency of the best whipped meringue! Traditionally, this dessert was beaten by hand or with a special stick with grass or strands of bark tied to one end. These tools were eventually replaced in more modern times by eggbeaters and mixers. Like egg whites, soapberries will not foam in plastic or greasy containers. The incredibly thick foam is rather bitter, so it was usually sweetened with sugar or with other berries. Soapberries can also be added to stews or cooked to make syrup, jelly, jam or a sauce for savoury meats. Canned soapberry juice, mixed with sugar and water, makes a refreshing "lemonade."

Beyond their bitter taste and soapiness, saponins can irritate the stomach and cause diarrhea or vomiting if consumed in large amounts so remember to eat soapberries in moderation.

Soapberries are rich in vitamin C and iron. They have been taken to treat flu and indigestion and have been made into a medicinal tea for relieving constipation. Other parts of the plant are used medicinally for various purposes, particularly by First Nations of western Canada where the species is more common. The berries can also be crushed or boiled to use as a liquid soap. Canned soapberry juice, mixed with sugar and water, was used to treat acne, boils, digestive problems and gallstones. Soapberry bark tea was traditionally used to treat eye troubles. The fruit was collected by beating the branches over a canvas or hide and then rolling the berries down a wooden board into a container to separate leaves and other debris.

EDIBILITY: edible

FRUIT: Yellow or right red, oval with a fine silvery scale, cherry-like achenes surrounded by juicy pulp.

SEASON: Flowers in April. Fruits ripen July to September.

DESCRIPTION: Deciduous shrub, open-formed, to 2 m tall, unarmed. Young twigs covered in a brown or rusty scale. Older twigs and branches brownish-red with orange flecks, sometimes fissured. Leaves somewhat thick, elliptic, smooth-edged, tip rounded, opposite, top green with short silvery scales, rusty underneath when young. Flowers yellowish to greenish, male and female flowers on separate plants, single or forming small clusters. Fruit grows on very short

Soapberry (*S. canadensis*)

stalks on female plants, at leaf axils. Grows in open woods and on streambanks and shores of western and southern Ontario, James Bay and Hudson Bay. Prefers moist habitat but will tolerate some drought.

Indian Ice Cream

Makes approximately 6 cups

Even with sugar this treat will have a slightly bitter taste, but many people quickly grow to like it.

1 cup soapberries • 1 cup water
4 Tbsp granulated white sugar

Put berries and water into a wide-topped ceramic or glass mixing bowl. *Do not use a plastic bowl or utensils, and make sure that nothing is greasy, or the berries will not whip properly.* Whip the mixture with an electric eggbeater or hand whisk until it reaches the consistency of beaten egg whites. Gradually add the sugar to the pink foam, but not too fast or the foam will "sink." Serve immediately.

Common Hackberry *Celtis occidentalis*

Also called: American hackberry, nettletree

Common hackberry (*C. occidentalis*)

While these berries can grow in great profusion and are sweet and tasty to eat, these qualities are off-set by the difficulty in climbing such a tall tree to pick the fruit and by the large size of the pit growing inside the relatively small drupes. Historically these fruits were used extensively by First Nations for food and also to season cooking meat. The hard berries can be ground and cooked to make a porridge or dried into cakes for winter consumption. They make nice jellies and preserves, but the amount of fin-ished produce is greatly diminished by the removal of the hard pits.

The wood of this relatively long-lived tree is light-coloured, yellowish-grey with brown and yellow streaks. However, it is not strong and rots easily, making it most appropriate for use in indoor furniture.

EDIBILITY: edible

FRUIT: Round purplish-black drupe to 11 mm in diameter, thin-fleshed, containing one large pit, growing singly or in pairs from the leaf axils on short stemlets.

SEASON: Flowers April to May. Fruits ripen August to September.

DESCRIPTION: Deciduous tree growing 13–20 m in height (but sometimes much larger under ideal conditions) and often an almost equal spread. Young branches tan and knobby with light-coloured lenticels. Mature bark dark grey, ridged, with a distinctive warty, cork-like texture. Leaves alternate on short stemlets (petioles), elongated heart-shaped, dull green, rough-textured, pale on underside. Leaf edges sharply toothed, but not at the base, which is smooth. Inhabits moist soils in valleys, bottomlands and hardwood forests and also does well in disturbed sites, field edges and urban areas. Prefers full sun and sometimes grows on drier sites.

WILD GARDENING: *Hackberry is a common planting in urban areas because of its rapid growth habit, decorative bark and ability to survive storm damage. It also provides important habitat and food for wildlife in urban areas. Dwarf hackberry (C. tenuifolia) is native to Ontario but only rarely found in the southern-most tip. It is more of a scraggly shrub than a tree and produces brick-red, dry but sweet fruits.*

Common hackberry (*C. occidentalis*)

Common hackberry (*C. occidentalis*)

Common hackberry (*C. occidentalis*)

143

Silverberry
Elaeagnus commutata

Also called: wolf willow

Silverberry (*E. commutata*)

The berries of this species are very dry and astringent, but some northern tribes gathered them for food. Most groups considered the mealy berries famine food and did not ingest them regularly. When eaten, they were consumed raw or cooked in soup. They were also cooked with animal blood, mixed with lard and eaten raw, fried in moose fat or frozen. Despite not being very palatable raw, the berries reportedly make good jams and jellies and are apparently much sweeter after exposure to freezing temperatures.

Some tribes used the nutlets inside the berries as decorative beads. The fruits were boiled to remove the flesh, and while the seeds were still soft, a hole was made through each. They were then threaded, dried, oiled and polished.

The flowers of silverberry can be detected from metres away by their sweet, heavy perfume. Some people enjoy this fragrance, but others find it overwhelming and nauseating. If green silverberry wood is burned in a fire, it gives off a strong smell of human excrement! Some practical jokers enjoy sneaking branches into the fire

Silverberry (*E. commutata*)

Silverberry (*E. commutata*)

and watching the reactions of fellow campers.

EDIBILITY: not palatable

FRUIT: Fruits silvery, mealy, about 1 cm long, drupe, with a single large nutlet.

SEASON: Flowers June to July. Fruits ripen late August to September.

DESCRIPTION: Thicket-forming, rhizomatous shrub to 3 m with 2–6 cm-long, alternate, lance-shaped, deciduous, silvery leaves covered in dense, tiny, star-shaped hairs (appearing silvery). Unarmed but young branches covered in orange-brown scales. Flowers strongly sweet-scented, yellow inside and silvery outside, 6–16 mm long, borne in two's or three's from leaf axils. Grows on well-drained or clayey shores and slopes, gravel bars and forest edges in northern and, occasionally, western Ontario.

145

Partridge Berry *Mitchella repens*

Also called: two-eyed berry

Partridge berry (*M. repens*)

The Mi'kmaq, Iroquois, Montagnais and Maliseet ate partridge berries fresh or preserved and sometimes cooked them into a jam. The Mi'kmaq also reportedly used this plant to make a beverage. The berries are considered edible, though not very tasty.

Historically, the whole plant, or more often the vine, was used medicinally. Some First Nations reportedly used partridge berry to ease childbirth. The Montagnais used a jelly made from the cooked berries to treat fevers, and the Abenaki made a salve of partridge berry mixed with plantain to reduce swelling.

The trailing stems and pretty red berries of this plant make a good Christmas decoration. Partridge berry was also combined with other plants and used for smoking. Wildlife such as ruffed grouse and wild turkey, and presumably partridge, enjoy the bud, leaf, flower and fruit of this plant.

EDIBILITY: edible

FRUIT: Fruits scarlet, double berries (ovaries of 2 flowers united) with indentation and distinctive 2 star-shaped marks, containing 8 seeds, persisting all winter.

SEASON: Flowers June to July. Fruits ripen August to September.

DESCRIPTION: Small, trailing, perennial, evergreen vine less than 50 cm long. Stems slender, wiry. Leaves dark green, opposite, stalked, blades blunt at tip, rounded at base,

Partridge berry (*M. repens*)

forming large mats. Found in southern Ontario's dry or moist mixedwoods, hardwood, cedar or pine forests as well as sandy to coarse, loamy, upland sites.

Partridge berry (*M. repens*)

1–2.5 cm long and wide, smooth, with a pale midrib, often variegated with white lines above; margins toothless. Flowers white (occasionally purple-tinged), tubular with coarse hairs inside, with usually 4 spreading lobes, 10–15 mm long, fragrant, in pairs. Roots easily, often at leaf nodes

Partridge berry (*M. repens*)

Yellow Clintonia   *Clintonia borealis*

Also called: yellow blue-bead lily

Yellow clintonia (*C. borealis*)

The unusual bead-like fruit of this species are variously reported as mildly toxic or poisonous so their consumption is not recommended. Young leaves of this species can be eaten raw or cooked and are said to taste slightly sweetish, like cucumber.

While the berries are mildly toxic to humans and were considered inedible by indigenous groups within their range, these plants nonetheless were used medicinally. A poultice of yellow clintonia leaves was applied to open wounds, infections, ulcers and bruises. A decoction of the plant was taken for the heart and to treat diabetes and was also used as a wash for the body. An infusion of the roots was used to aid childbirth. The crushed leaves of yellow clintonia were rubbed on the face and hands as a protection from mosquitoes.

EDIBILITY: edible with caution (toxic), poisonous

FRUIT: Berries round, in clusters, bright metallic blue, 6–12 mm thick.

SEASON: Flowers May to June. Fruits ripen in August.

DESCRIPTION: Perennial herb arising from rhizomes, usually growing in patches. Leaves basal, 2–4 but normally 3, dark green, glossy, thick and leathery, simple, up to 30 cm long and 9 cm wide, narrowed at both ends. Flowers of 6 greenish-yellow petals, cup-shaped, 1–3 cm long. Flowering stem 14–40 cm tall, usually hairy at the top, flowers in clusters of 2–8. Found in all soil types and moisture regimes in Ontario, particularly mixed and boreal forests.

WARNING: *Be careful not to confuse these mildly toxic berries for much tastier fruits of similar appearance—such as blueberries.*

Yellow clintonia (*C. borealis*)

Yellow clintonia (*C. borealis*)

Yellow clintonia (*C. borealis*)

Yellow clintonia (*C. borealis*)

Twisted-stalks *Streptopus* spp.

Rosy twisted-stalk (*S. lanceolatus*)

These perennials are called "twist-ed-stalks" because of the kinks (sometimes right-angled, sometimes just a sharp curve) present in the main stem or flower stalks. Most Native peoples regarded twisted-stalks as poisonous and used the plant mainly for medicine, but some ate young plants and/or the bright-coloured berries, either raw or cooked in soups and stews. The berries are juicy and moderately sweet-tasting.

Twisted-stalks were highly regarded for their general restorative qualities and were taken as a tonic or to treat general sickness. Some First Nations used the plant to treat coughs, loss of appetite, stomachaches, spitting up blood, kidney trouble and gonorrhea. The blossoms were ingested to induce sweating. The plant was sometimes tied to, and used to scent, the body, clothes or hair. Twisted-stalks differ from the closely related fairybells in that the flowers attach to the stem in the leaf axils instead of to the branch tip.

EDIBILITY: edible (with caution)

FRUIT: Berries hanging, red-orange or yellowish, egg-shaped and somewhat translucent; seeds small, whitish, somewhat visible.

SEASON: Flowers late June to early August. Fruits ripen August to September.

DESCRIPTION: Slender, herbaceous perennial, from thick, short rhizomes, 0.4–1 m tall or more, with characteristic stems zig-zagging or curved from leaf to leaf, occasionally branched. Leaves smooth-edged, elliptical/oval-shaped, alternate, markedly parallel-veined. Flowers small, white, bell-shaped, 8–12 mm long, with 6 petals that flare backward, each hanging on the lower side of each stalk, 1 per leaf. Found in moist, shaded forests, thickets, clearings, meadows, disturbed sites and on streambanks.

WARNING: *Young twisted-stalk plants closely resemble green false-hellebore, which is extremely poisonous. Collecting the young shoots of twisted-stalk for consumption is not recommended unless you are absolutely sure of plant identification!*

Rosy twisted-stalk (*S. lanceolatus*)

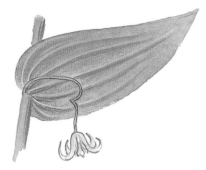

Clasping twisted-stalk (*S. amplexifolius*)

Rosy twisted-stalk (*S. lanceolatus*)

Clasping twisted-stalk (*S. amplexifolius*)

Rosy twisted-stalk (*S. lanceolatus*)

Clasping twisted-stalk (*S. amplexifolius*)

Clasping twisted-stalk (*S. amplexifolius*)

Clasping twisted-stalk (*S. amplexifolius*) grows 0.5–1 m tall and has branched, smooth stems, bent at nodes giving it a zig-zag appearance. Leaves smooth, heart-shaped, clasping at base. Flowers greenish-white. Berries bright red. Grows in the deciduous forests of southern Ontario north and east into the boreal forest. Also called: clasping-leaved twisted-stalk, white mandarin, liverberry.

Rosy twisted-stalk (*S. lanceolatus*) grows to 30 cm tall, with stems usually unbranched, curved (not zig-zagged), leaves not clasping and with fine hairs along the margins. Rose-purple or pink flowers with white tips. Red berries. Found throughout most of Ontario. Also called: syn. *S. roseus*.

Clasping twisted-stalk (*S. amplexifolius*)

Canada Mayflower

Maianthemum canadense

Also called: wild lily-of-the-valley

Canada mayflower (*M. canadense*)

Berries of this species are considered edible but are bitter-tasting and not very palatable. This species has a wide range across Canada, and many First Nations consumed the fruit, but they were rarely highly regarded. The berries were usually only eaten casually by children, or by hunters and berry pickers while out on trips.

This plant was used to treat headaches and sore throats and "to keep kidneys open during pregnancy." A leaf poultice was used to treat swellings in the limbs. Canada mayflower was also used to make smoke for inhaling. The genus name for this plant is derived from the Latin word for "May," referring to the flowering time of these plants.

EDIBILITY: not palatable

FRUIT: A true "berry," pea-sized, at first hard and green, maturing to cream-coloured with red speckles, then pinkish with red flecks, ripening to a solid red.

SEASON: Flowers May to June. Fruits ripen July to September.

DESCRIPTION: Small, herbaceous, creeping perennial herb usually growing to less than 25 cm tall, arising from rhizomes and usually forming large colonies. Leaves stalk-less, heart-shaped, alternating, usually 2 or 3, with prominent parallel veins.

Flowers small, white, with 4 petals, 4–6 mm wide, borne in distinct terminal clusters, blooming early spring. Fruit borne at the top of stems, in clusters. Found in moist woods and clearings, often abundant in all but the far north.

WILD GARDENING: *This species grows into a delightful, dense carpet of delicate heart-shaped leaves and makes an excellent low-maintenance understorey planting for the woodland garden. The flowers are pretty in spring, and the berries provide a showy late-summer and fall display that attracts wildlife such as grouse that like to eat them.*

Canada mayflower (*M. canadense*)

Canada mayflower (*M. canadense*)

Canada mayflower (*M. canadense*)

155

False Solomon's-seals *Maianthemum* spp.

Common false Solomon's-seal (*M. racemosum*)

Various indigenous peoples across Canada ate the ripe berries, young greens and fleshy rhizomes of this genus but not universally as these foods were not generally highly regarded. In cases where berries were eaten, it was usually taken casually (by hunters, berry pickers, children). The Ojibwa soaked the rhizomes in lye "to get rid of their disagreeable taste" and then cooked them like potatoes.

False Solomon's-seal was often combined with other plants for medicinal purposes. When combined with dogbane, it was used to keep the kidneys open during pregnancy, to cure sore throats and headaches and as a reviver. When mixed with black ash the plant was used to loosen the bowels. When mixed with sweetflag, false Solomon's-seal was used as a conjurer's root to perform tricks or cast spells. Both the leaf and root were used to reduce bleeding. The rhizome was burned and inhaled to treat a number of ailments: to relieve head-aches, to quiet a crying child and to return someone to normal after temporary insanity. Leaf decoctions were applied to assist in childbirth and to treat itchy rashes. A rhizome decoction was used to treat back pain and overexertion. Berries of this species can cause diarrhea so should be eaten with caution.

EDIBILITY: not palatable

FRUIT: See individual species descriptions.

SEASON: Flowers May to June. Fruits ripen August to October.

DESCRIPTION: Herbaceous perennials growing from thick, whitish, branching rhizomes, often found in dense clusters. Leaves smooth-edged, broad, elliptical, alternate along stems in 2 rows, 5–15 cm long, distinctly parallel-veined, often clasping. Flowers small, cream-coloured, 6-parted, in dense, terminal clusters. Berries small and densely clustered, initially green and mottled or striped, ripening to bright red. Grows in rich woods, thickets and moist or wet clearings across Ontario.

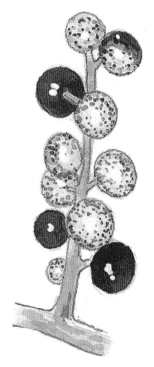

Common false Solomon's-seal (*M. racemosum*)

Star-flowered false Solomon's-seal (*M. stellatum*)

Common false Solomon's-seal (*M. racemosum*) grows in clumps to 1.2 m tall from a fleshy, stout rootstock. Stems unbranching, arching, in a zig-zag form. Leaves lance-shaped, bright green. Flowers in pyramid-like clusters of 50–70 producing a tight cluster of many seedy berries. Berries at first green with copper-coloured spots, ripening to red, often with purple spots. Found across Ontario except in the Hudson Bay Lowlands. Also called: feathery false lily-of-the-valley, false spikenard • *Smilacina racemosa*.

Star-flowered false Solomon's-seal (*M. stellatum*) grows to 60 cm tall, flowers in clusters, berries at first green with blue-purple stripes. Differentiated from *M. racemosum* by being smaller, with fewer flowers and leaves and fewer berries (2–8) that are larger and green with red stripes when unripe. Also called: starry false lily-of-the-valley, little false Solomon's-seal • *Smilacina stellata*.

Three-leaved Solomon's-seal (*M. trifolium*) is a slender, erect plant growing to 20 cm tall. Leaves 2–4 (usually 3), 5–12 cm long, 1–5 cm wide, stalkless, clasping or sheathing the stem or base, smooth, ascending. Berries round, dark red. Differentiated from Canada mayflower by having longer, more lance-shaped leaves with no stalks. Inhabits sphagnum bogs and wet conifer swamps. Also called: three-leaved smilacina • *Smilacina trifolia*.

Star-flowered false Solomon's-seal (*M. stellatum*)

Star-flowered false Solomon's-seal (*M. stellatum*)

Three-leaved Solomon's-seal (*M. trifolium*)

Common false Solomon's-seal (*M. racemosum*)

Common false Solomon's-seal (*M. racemosum*)

Three-leaved Solomon's-seal (*M. trifolium*)

Three-leaved Solomon's-seal (*M. trifolium*)

Rough-fruited fairybells (*P. trachycarpa*)

These berries were not widely eaten by First Nations in Canada, and although some tribes such as the Shuswap and Blackfoot reportedly ate the fruits, other groups considered them poisonous. A compound decoction of roots was taken for internal pains, and an infusion of leaves was used to clean wounds. The whole plant can be boiled and the water drunk as a spring tonic.

EDIBILITY: edible with caution (toxic)

FRUIT: Berries orange to bright red, conspicuously rough-skinned, velvety-surfaced with 6–12 seeds.

SEASON: Flowers April to July. Fruits ripen in July.

DESCRIPTION: Perennial herb, growing 30–80 cm tall, with few branches, from thick-spreading rhizomes. Leaves alternate, broadly oval, 3–9 cm long, pointed at tips, rounded to a heart-shaped base, fringed with short, spreading hairs, prominently parallel-veined. Flowers creamy to greenish white, narrowly bell-shaped, 10–20 mm long, drooping, 1–3 at branch tips. Grows in rich soils in shady deciduous forests, open coniferous forests and in moist woods and thickets.

Rough-fruited fairybells (*P. trachycarpa*)

Rough-fruited fairybells (*P. trachycarpa*)

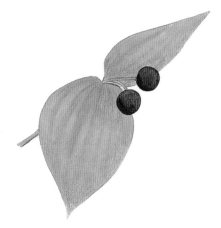

Rough-fruited fairybells (*P. trachycarpa*)

Rough-fruited fairybells (*P. trachycarpa*)

Indian cucumber root (*M. virginiana*)

The fleshy rhizome of this plant is edible and has a pleasant cucumber-like taste. The dark purple berries, however, are considered inedible and should not be consumed. The Iroquois reportedly used the dried berries and leaves of this plant as an anticonvulsive for babies.

EDIBILITY: inedible

FRUIT: Dark purple glossy berries, 5–14 mm in diameter, clustered.

SEASON: Flowers June to July. Fruits ripen August to September.

DESCRIPTION: Perennial herb growing on single stems 20–90 cm, in colonies, from a fleshy rhizome. Each stem with two whorls of leaves; a whorl of 6–10 lance-shaped leaves to 12 cm long at the bottom and a second whorl of 3 or 4 broader leaves, 2.5–7.5 cm long, mid-way up the stem. Leaves parallel-veined, smooth-edged, shiny, tapering. Flowers clustered in group of 3–9, hanging, yellowish-green. Fruits clustered above the top whorl, on short red stemlets. Inhabits open semi-shaded areas in moist coniferous, deciduous and mixedwood forests, swamp and bog edges in southern Ontario.

Indian cucumber root (*M. virginiana*)

Indian cucumber root (*M. virginiana*)

Indian cucumber root (*M. virginiana*)

Solomon's-seals   *Polygonatum* spp.

Smooth Solomon's-seals (*P. biflorum*)

The berries of this genus are considered toxic and inedible, although the Iroquois around Lake Ontario reportedly used a dried powder of smooth Solomon's-seal roots in bread. The berries are said to taste mucilaginous, sweet, then acrid. First Nations ate the roots of hairy Solomon's-seal either raw or cooked, and also made a flour from them when dried. A root tea was used to treat coughs. The rhizomes have a history of medicinal use in the treatment of skin discoloration, ulcers and broken bones.

This is a very decorative and charming species, with its delicate arching stems, demure blooms, striking berries, and fine foliage that turns a lovely yellow in the fall. It makes a great addition to the shade or woodland garden, and once its tough rhizomes are established, it slowly spreads to create a drought-resistant patch. This plant is a great attraction for hummingbirds, which feed on the flower nectar.

Hairy Solomon's-seals (*P. pubescens*)

Smooth Solomon's-seal is the largest of this plant group and is often found in ornamental gardens because of its great size and beauty.

EDIBILITY: edible with caution (toxic)

FRUIT: Dark-blue to black berries hang down from leaf axils. See individual species descriptions for more specific information.

SEASON: Flowers April to June. Fruits ripen July to August.

DESCRIPTION: Herbaceous perennials growing on single arching stems from a thick, fleshy, many-jointed white rhizome. Often spreads into colonies. Leaves, simple, stalkless, finely veined, alternate, opposite in two close ranks. Flowers bell-shaped, tubular, 7–20 mm long, borne below arching stems from leaf axils. Inhabits forests and woodlands, preferring shade or partial sun in light, well-drained, humus-rich soil.

Hairy Solomon's-seal (*P. pubescens*) grows to 1 m tall. Leaves broadly oval to lance-shaped, 5–15 cm long, 1.5–7.5 cm wide, fine hairs along the 3–9 prominent veins. Flowers greenish to white, 7–13 mm long, usually hanging in pairs. Berries dark blue to black, often in pairs, several-seeded. Grows in southern and western Ontario in both dry and moist forests.

Smooth Solomon's-seal (*P. biflorum*) grows 0.5–2 metres tall and generally 45–60 cm wide. Leaves smooth, linear to ovate with 1–5 prominent veins. Flowers bright or greenish white, up to 2 cm long and normally in clusters of 2–7 in early to late spring. Berries round, dark blue, 1 cm in diameter, with a whitish bloom, growing in small groups or singly, each with a thin vertical cleft. Common only south of the Canadian Shield. Also called: great Solomon's-seal.

Smooth Solomon's-seals (*P. biflorum*)

Hairy Solomon's-seals (*P. pubescens*)

Strawberries *Fragaria* spp.

Woodland strawberry (*F. vesca*)

These delicious little berries pack significantly more flavour than a typical large, domestic strawberry. Wild strawberries are small compared to modern cultivars and are probably best enjoyed as a nibble along the trail, but they can also be collected for use in desserts and beverages. A handful of bruised berries or leaves, steeped in hot water, makes a delicious tea, served either hot or cold.

Strawberries were popular with all First Nations, but the berries' juiciness can make them difficult to dry and preserve. Today, strawberries are prepared by freezing, canning or making jam, but traditionally, they were mostly eaten fresh or sometimes sun-dried. The berries were mashed and spread over grass or mats to dry in cakes, which were later eaten dry like fruit leather or rehydrated, either alone or mixed with other foods as a sweetener. Anyone who has had the extreme pleasure of savouring dried wild strawberries knows that they are a treat well worth the time to prepare! Strawberry flowers, leaves and stems were sometimes mixed with roots in cooking pits as a flavouring.

Strawberries contain many quickly available minerals (e.g., sodium, calcium, potassium, iron, sulphur and silicon), as well as citric and malic acids, and they were traditionally used to enrich the bloodstream. Strawberry leaf tea, accompanied by fresh strawberries, was a recommended

Woodland strawberry (*F. vesca*)

Wild strawberry (*F. virginiana*)

remedy for gout, rheumatism, inflamed mucous membranes and liver, kidney and gallbladder problems. Strawberries are a good source of ellagic acid, a chemical that is believed to prevent cancer and appears to support healthy aging of the brain. To remove tartar and whiten discoloured teeth, strawberry juice can be held in the mouth for a few minutes and then rinsed off with warm water. This treatment is reported to be most effective with a pinch of baking soda in the water. Large amounts of this fruit in the diet also appear to slow dental plaque formation. Strawberry juice, rubbed into the skin and later rinsed off with warm water, has been used to soothe and heal sunburn.

Wild strawberry and coastal strawberry (*F. chiloensis*, native to coastal

BC) are the original parents of 90 percent of our modern cultivated strawberry varieties.

EDIBILITY: highly edible

FRUIT: "Berries" (technically, a receptable with achenes) are red when ripe, resembling miniature cultivated strawberries.

SEASON: Flowers May to August. Fruits ripen in June, and flowers continue to bloom throughout summer, so plants often have ripe fruit and flowers on them at the same time.

DESCRIPTION: Low-creeping perennials with long, slender runners (*stolons*). Leaves green, often turning red in fall, 5–10 cm wide, with 3 sharply toothed leaflets on hairy stalks. Flowers white, 5-petalled, 1.5–2 cm across, usually several per stem, forming small, loose clusters.

Wild strawberry (*F. virginiana*) has bluish-green leaflets, with the end tooth narrower and shorter than its adjacent teeth. Common throughout Ontario in dry to moist open woodlands and clearings, often in disturbed areas on well-drained sites, but growing in a wide range of habitat. Also called: common strawberry.

Woodland strawberry (*F. vesca*) has yellowish-green leaflets, with the end tooth projecting beyond its adjacent teeth. Leaflets are thick, hairy, strongly veined and scalloped. Flowering stems taller than leaves. Inhabits dry to moist soils in sunny woodlands and forests and along edge habitat and disturbed areas such as roadsides and lakeshores throughout Ontario.

Wild strawberry (*F. virginiana*)

Woodland strawberry (*F. vesca*)

INTERESTING: *Many people will be surprised to learn that the strawberry is technically not a fruit! What we think of as the "berry" is actually a swollen receptacle (this is the base of the flower, which you would normally expect to see inside a fruit). The true "fruits" are the tiny dark seeds (seed-like achenes) that you can easily find either embedded in (wild strawberry), or perched on (woodland strawberry), the soft flesh of the strawberry.*

Wild strawberry (*F. virginiana*)

Wild strawberry (*F. virginiana*)

Wild strawberry (*F. virginiana*)

Wild Berry Muffins

Makes 12 muffins

This batter can also be baked in a loaf form.

5 Tbsp vegetable oil • 2 eggs, lightly beaten
1½ cups mixed wild berries (strawberries, thimbleberries, blueberries, huckleberries, dewberries, etc.)
1 tsp salt • 1¾ cups whole wheat flour
¾ cup brown sugar • 2¼ tsp baking powder

Preheat oven to 400° F. Mix wet ingredients together in a bowl. Sift dry ingredients together in another bowl. Make a shallow well in the centre of the dry ingredients and slowly add the wet mixture. Mix well and pour into greased or lined muffin tins. Bake for 10 to 15 minutes, or until a knife inserted into a muffin comes out clean.

Ginseng *Panax* spp.

Dwarf ginseng (*P. trifolius*)

More famous for their aromatic roots, which can be eaten raw, cooked or candied and chewed, ginseng berries are also edible and taste much like the root. The leaves can be used to make a pleasant tea. The most popular and best documented use of ginseng root is as an adaptogen, a substance with the ability to boost the immune system, increase mental efficiency and physical performance and aid in adapting to high or low temperatures and stress. The North American species are considered to have similar medicinal properties to the Asian species, *P. ginseng*, though they are milder in their potency and are thus prescribed primarily to younger patients.

Ginseng root has a rich 5000-year history of herbal use and plays a crucial role in regulating the balance between yin and yang according to traditional Chinese medical theory. In China, the root is highly prized as an aphrodisiac

and as a panacea to promote health, vigour and long life. In the West, ginseng is used as a general health tonic useful for promoting appetite and digestion. Considered a virtual cure-all, the plant is prescribed for a wide array of disorders, including stomach aches, neuralgia, rheumatism, gout, irritation of the respiratory tract, gastrointestinal illness, weak circulation and a variety of nervous disorders.

Research has shown that ginseng both stimulates and relaxes the nervous system, promotes hormone secretion and improves stamina and disease resistance. Its ability to lower blood sugar and cholesterol levels is being studied, and the herb may one day provide a therapeutic benefit to diabetic patients. Isolated ginseng saponins have demonstrated a strong radio-protective effect in cancer patients prior to gamma-irradiation. In Canada, First Nations used a decoction of ginseng root for a variety of ailments, including fevers, coughs, headaches, nausea and vomiting. Although the herb was not normally prescribed to pregnant women according to Chinese practice, in North America, women regularly consumed a decoction of the root for everything from normalizing menstruation to easing childbirth. The age of the root is believed to correspond to its potency, and older roots should theoretically be taken in smaller doses. Some First Nations have used ginseng as an ingredient in love potions and charms.

Although dwarf ginseng roots are not commercially prized, they are edible and were eaten by some First Nations as a starch source and used medicinally by the Iroquois and Cherokee south of the border.

EDIBILITY: edible

FRUIT: Red or yellow berries with 2–3 seeds.

SEASON: See individual descriptions.

DESCRIPTION: Perennial herb from a fleshy, often forked taproot. Tiny flowers with 5 petals grow in round clusters atop a slender stalk. Fruits berries.

WILD HARVEST RESTRICTIONS: *Commercial exploitation in Canada during the 1700s has significantly decimated the ginseng population such that it is now illegal to export wild plants and roots. Canada has since become the world's largest commercial grower of cultivated American ginseng, and it is Ontario's fifth most valuable cash crop. Considered a species at risk in Ontario, ginseng is protected and a hefty fine is imposed for collection or distribution of wild specimens.*

American ginseng (*P. quinquefolius*) is 30–50 cm tall, with whorls of 3 long-stalked leaves, 12–30 cm long, each leaf divided palmately into usually 5 sharp-toothed leaflets on stalks. Flowers whitish or yellow-green, 2 mm wide, blooming from July to August. Fruit a 2-seeded, red berry to 12 mm across, growing in clusters. Grows in moist, rich woods in southern Ontario.

Dwarf ginseng (*P. trifolius*) is small, 10–20 cm tall, with whorls of 3 leaves, each palmately divided into 3–5 coarsely-toothed leaflets, to 8 cm long and stalkless. White flowers appear from May to June. Fruit a yellow, 2- to 3-seeded berry of about 5 mm. Grows in rich woodlands in southern Ontario.

American ginseng (*P. quinquefolius*)

American ginseng (*P. quinquefolius*)

Sarsaparillas *Aralia* spp.

Wild sarsaparilla (*A. nudicaulis*)

These fragrant plants (which belong to the ginseng family) have a warm, aromatic, sweetish taste that is most intense in the rhizomes and berries. The berries were used to flavour beer and to make wine (similar to elderberry wine), and a tea was sometimes made from the seeds. The berries are generally considered inedible, and while some sources recommend them as a flavouring, others report them as mildly toxic. Some people use the berries to make jelly but this is not recommended (see Warning). The rhizomes were generally considered emergency food only, but some indigenous hunters and warriors are said to have subsisted on them during long trips. The rhizomes were traditionally used to make tea, root beer and mead. Young shoots were sometimes also cooked as a potherb.

The rhizomes (and occasionally the leaves) were pulverized by pounding or chewing and were used in poultices to soothe and heal wounds, burns, sores, boils and other skin problems and to relieve swelling and rheumatism. Mashed rhizomes were also stuffed into noses to stop bleeding and into ears to stop aching. The rhizomes and berries were boiled to make medicinal teas and syrups or soaked in alcohol to make tinctures. These medicines were used to treat many different problems, ranging from stomachaches to rheumatism and syphilis. The pleasant-tasting rhizome tea was valued as a blood-purifier, tonic and stimulant and as a medicine for stimulating sweating. It was also used for treating lethargy,

Bristly sarsaparilla (*A. hispida*)

general weakness, stomachaches, fevers and coughs. Wild sarsaparilla was widely used in patent medicines in the late 1800s.

EDIBILITY: edible with caution (toxic)

FRUIT: Dark berries.

SEASON: See individual species descriptions.

DESCRIPTION: Perennial shrubs or herbs growing from rhizomes. Leaves large and compound. Flowers 5-parted, whitish and growing in terminal clusters.

Bristly sarsaparilla (*A. hispida*) is a shrub to 1 m tall, with sharp, stiff bristles at the base of each tall erect stem. Leaves twice compound with oval or lance-like, toothed leaflets. Flowers small, greenish white and growing in globe-shaped clusters at the top of stems, from June to August. Fruits dark purple to black, foul-smelling berries. Grows frequently in sandy, open woods and rocky areas or clearings throughout the province but less common in the southern- and northern-most regions. Also called: hairy sarsaparilla, dwarf elder.

Spikenard (*A. racemosa*) is a perennial herb, to 2 m tall, with dark-green or reddish stems. Leaves 3 times compound with 6–21 toothed, heart-shaped leaflets, each up to 15 cm long with toothed margins. Flowers small, whitish, growing in small clusters along a branching raceme, from June to August. Grows in rich, moist woods and swamps across the southern parts of the province.

Wild sarsaparilla (*A. nudicaulis*) is a herb, to 70 cm tall but normally shorter, with a long, horizontal rhizome. Leaf blades horizontal, with 3 major divisions, each of these divided into 3–5 oval leaflets 3–12 cm long. Flowers usually hidden under the leaf, greenish white, 5–6 mm long, forming 2–7 (usually 3) round, 2–5 cm wide clusters, from May to June. Fruits dark purple berries, 6–8 mm long, ripening July to August. Grows abundantly in a wide range of soil moisture and types throughout Ontario.

Wild sarsaparilla (*A. nudicaulis*)

Spikenard (*A. racemosa*)

WARNING: *Some people have reported being very sick after eating wild sarsaparilla berries.*

Bunchberry *Cornus canadensis*

Also called: Canada dogwood, dwarf dogwood • *C. unalaschensis*

Bunchberry (*C. canadensis*)

The bright scarlet-orange fruits of this woodland plant look like they should be very poisonous but are actually quite edible. However, opinions of their flavour range from insipid to a pulpy, sweet, flavourful fruit. Many First Nations within the extensive range of this plant gathered these fruit in quantity. They were eaten fresh as a snack or gathered and stored for winter use. In more modern times they are enjoyed with granulated sugar.

The berries are abundant where the plant grows, and it is easy to gather a basketfull with minimal effort. They can be eaten raw as a trail nibble and are also said to be good cooked in puddings. However, each drupe contains a seed that is quite hard, so be wary if you have dental work. Bunchberries (often mixed with other fruits) can be used whole to make sauces and preserves or cooked and strained to make beautiful scarlet-coloured syrups and jellies.

The berries are reported to have anti-inflammatory, fever-reducing and pain-killing properties (rather like mild aspirin), but without the stomach irritation and potential allergic effects of salicylates. The plant has a history of being used to treat headaches, fevers, diarrhea, dysentery and inflammation of the stomach or large intestine. The berries were eaten and/or applied in poultices to reduce the potency of certain poisons. They were also chewed and the resulting pulp applied topically to soothe and treat burns. Bunchberries have historically been steeped in hot water to make a medicinal tea for treating paralysis, or boiled with tannin-rich plants (such as common bearberry or commercial tea) to make a wash for relieving bee stings and poison-ivy rash. Native peoples used tea made with the entire plant to treat aches and pains, lung and kidney problems, coughs, fevers and fits.

EDIBILITY: edible

FRUIT: Bright orange-red berry-like drupes, 6–9 mm wide, growing in dense clusters at the stem tips, nestled into a whorl of leaves (hence the common name "bunchberry"). The drupe has a yellowish pulp and single seed.

SEASON: Flowers May to August. Fruits ripen July to August.

Bunchberry (*C. canadensis*)

DESCRIPTION: Perennial, rhizomatous herb, 5–20 cm tall. Leaves 2–8 cm long, pointed at both ends, growing opposite each other in groupings of 2–6, spaced so tightly that they have a whorled appearance. Flowers tiny, in a dense clump at the centre of 4 white to purple-tinged, petal-like bracts (exactly like miniature flowers of the dogwood tree), forming single, flower-like clusters about 3 cm across. Grows throughout Ontario in cool, moist woods and damp clearings at low to subalpine elevations, commonly found on rotting stumps and logs.

Bunchberry (*C. canadensis*)

AMAZING: *This plant spreads its relatively heavy pollen grains through an interesting "explosive pollination mechanism." When the pollen is ripe and ready to be released, an antenna-like trigger lets go in the flower, rapidly springing the four pollen-laden anthers violently upward together in a snapping motion, thereby catapulting the pollen grains far up into the air for dispersal.*

Ground Cherries
Physalis and *Leucophysalis* spp.

Also called: cape gooseberry

Clammy ground-cherry (*P. heterophylla*)

While the plants themselves are poisonous, the fruit of ground cherries are delicious, particularly the clammy ground-cherry, Ontario's most abundant species in the group. An orange jewel held in a dry, papery husk, the fruit pops like a sweet-sour explosion in the mouth when eaten. Equally tasty fresh, in pies or in jam, it also stores well for months in the fridge or a cool room, providing a welcome burst of summer flavour as late as Christmas. Although this plant is native to eastern Canada, it grows well in warm, sunny gardens across the country. Many eastern First Nations relished the berries as a fresh autumn food source or dried them for winter use. The Iroquois used an infusion of dried leaves and rhizomes externally as a healing wash for scalds and burns and to treat sores caused by venereal disease, and internally as an emetic. A poultice of rhizomes and leaves was used to treat wounds and inflammation. Ground-cherries do well in both the ornamental and vegetable garden.

EDIBILITY: highly edible

FRUIT: Fruit a many-seeded orange (when ripe) fruit 3 cm long, 2.5 cm in diameter, enclosed in a papery husk (*calyx*) that starts out green and becomes light brown when the fruit is ripe.

SEASON: Flowers in June and continue into fall. Fruits ripen August to September and into October in mild years.

DESCRIPTION: Perennial plants generally less than 1 m in height with erect, branching stems, hairy or smooth, growing from rhizomes. Leaves alternate, oval to elliptical or heart-shaped with a few coarse, irregular teeth on margins. Single, yellow or white, funnel-shaped flowers hang from leaf axils, to form the characteristic, calyx-enclosed fruit. Grows in fields and open woods.

Clammy ground-cherry (*P. heterophylla*) grows to 90 cm tall from stout rhizomes. Stems erect, densely hairy, branching. Leaves glandular-hairy on upper and lower surfaces, 6 cm long and 5 cm wide, alternate, oval, typically heart-shaped with a few coarse, irregular teeth on margins. Single, yellow, funnel-shaped flowers to 2 cm across hang from leaf axils, purplish at base, glandular-hairy externally. Yellow berries about 13 mm in diameter enclosed by a green calyx up to 4 cm long. Grows in fields and open woods of southern Ontario.

Smooth ground-cherry (*P. virginiana*) is similar to clammy ground cherry but generally shorter (to 60 cm) and relatively hairless. Most commonly found (as a pest) growing in cultivated fields of southern Ontario. Also called: Virginia ground-cherry.

White ground-cherry (*L. grandiflora*) is a bushy herb 30–50 cm tall with hairy glandular stems, smooth oval leaves to 10 cm long and yellowish or white funnel-shaped flowers up to 6 cm across that continue to bloom from May through August. Uncommon in southwest Ontario into the Great Lakes region in sandy, gravelly, or rocky disturbed sites. Also called: large white, or large-flowered, ground-cherry.

WARNING: *All parts of these plants are poisonous, including the unripe green fruit, so make sure that fruit is ripe and orange before ingesting it. Green fruit can ripen well off the plant (indeed, the husks often fall to the ground while the fruit is still green) and are not toxic when fully ripened. Ground-cherries are similar in appearance to the botanically-related tomatillo popular in Mexican cuisine, but tomatillos do not grow in the wild in Ontario.*

Clammy ground-cherry (*P. heterophylla*)

Smooth ground-cherry (*P. virginiana*)

Northern comandra (*G. lividum*)

The berries of this Canada-wide species are edible, but reports vary considerably regarding their tastiness. Northern comandra fruit were eaten by the Fisherman Slave Lake tribe but were considered inedible by most First Nations within the plant's range and are probably best used as an emergency food if lost in the woods. This species is parasitic, feeding off the roots of host plants.

EDIBILITY: not palatable

FRUIT: Fleshy orange to scarlet berry-like drupe 6–10 mm.

SEASON: Flowers May to June. Fruits ripen August to September.

DESCRIPTION: Perennial deciduous herb from reddish creeping underground rhizomes growing 10–25 cm tall. Flowering stems erect, unbranched. Leaves simple, alternate, short-stalked, 1–2.5 cm long and 0.5–1 cm wide, pale grey to purplish tinged, toothless margins. Flowers grow in clusters of 2–4 from leaf axils, green to purplish, 5 petal-like sepals. Fruits single, rarely double. Inhabits moist coniferous forests of northern Ontario and shorelines, and rocky woodlands and outcrops farther south.

Northern comandra (*G. lividum*)

Winterberry *Ilex verticillata*

Also called: American winterberry, black alder, Canada holly, fever bush

Winterberry (*I. verticillata*)

The berries of this native species have been prescribed as a vermifuge, and the bark (once dried and aged) was historically used as a laxative. However, without this aging, the bark (both the outer and inner layers were used) induces strong vomiting and stomach pain. A decoction of the powdered bark has been used as an external wash to treat gangrene, ulcers and eruptive skin conditions. The bark of this species is high in tannin and therefore has astringent properties.

Winterberry prefers wetlands but also does well in dry soils. It is hardy and disease resistant and can provide a decorative horticultural display year-round. The berries often persist on the tree throughout winter, thereby providing important food for birds and mammals. The bare branches covered in bright red berries are an excellent and striking addition to winter floral arrangements.

EDIBILITY: poisonous

FRUIT: Globose red drupe 6–8 mm, containing 3–5 bony nutlets, growing singly or in bunches of 2–3, persisting on the tree into winter.

SEASON: Flowers May to July. Fruits ripen August to September

DESCRIPTION: Deciduous erect shrub growing 1–5 m tall. Bark greyish, smooth, with lenticels.

Branchlets hairless, finely ridged, bark brown maturing to grey or blackish with lenticels. Leaves alternate, simple, 3.5–9 cm long and 1–4 cm wide, glossy to dull green above, hairy or hairless below, thick, serrated margins, growing on grooved hairy 1 cm stalks. Flowers small, greenish- to yellowish-white, 5 mm, 5–8 petals. Male flowers in clusters of 2–10, females in clusters of 1–3. Dioecious (separate female and male trees). Inhabits wetlands, swampy woods and damp thickets, uncommon in grasslands and drier areas. Abundant across southern Ontario with a few populations west of Lake Superior.

Winterberry (*I. verticillata*)

Winterberry (*I. verticillata*)

Winterberry (*I. verticillata*)

Mountain Holly *Nemopanthus mucronatus*

Also called: catberry • *Ilex mucronata*

Mountain holly (*N. mucronatus*)

While some reports list the berries of this species as edible, others claim the fruit is poisonous. In any case, the berries are reportedly strong and bitter-tasting so are not palatable for modern tastes. The fruits were used medicinally by First Nations in a number of ways. The Malecite made a medicine for coughs, fevers and tuberculosis by mixing mountain holly with blackberry roots, stahorn sumac, lily roots and mountain raspberry roots. A decoction of the branches was also taken as a general tonic.

EDIBILITY: edible with caution (toxic), poisonous

FRUIT: Round, red to purplish berry-like drupe, 6 mm in diameter, on slender stalks, contains 4–5 nutlets.

SEASON: Flowers in late May. Fruits ripen August to September.

DESCRIPTION: Erect, heavily

branched deciduous shrub growing 0.3–3 m tall. Bark grey, ashy, mostly smooth with many lenticels. Branchlets terminal, slender, purplish. Leaves alternate, bright green above, paler below, 7 cm long, 2.5 cm wide, on slender purplish stalks to 1 cm. Leaves on branchlets distantly spaced, on lateral branchlets crowded, appearing whorled. Flowers separate male and female on same plant. Drupes grow on short stems from the leaf axils, singly or 2–3 together, persisting throughout winter. Inhabits wet organic sites such as swamps, pond edges and damp thickets from the St. Lawrence River west to the eastern shores of Lake Superior.

Mountain holly (*N. mucronatus*)

Mountain holly (*N. mucronatus*)

Mountain holly (*N. mucronatus*)

183

Buckthorns · *Rhamnus* spp.

European buckthorn (*R. cathartica*)

Some sources report that the purple berries of these small trees or shrubs are edible, but possibly in modest amounts given their strong purgative effects. Glossy buckthorn and European buckthorn are introduced species that were brought to North America as ornamental and hedging plants and also as windbreaks. Both species have escaped cultivation and have become naturalized in some areas of southern Ontario, particularly around urban areas. Indeed, these plants have reproduced so successfully that they are now considered an invasive species and a threat to native plants in some regions.

Glossy buckthorn has traditionally been prescribed as a laxative. The green fruit produces a good green dye and the bark a shade of yellow. The bark, leaves and fruit of this species contain strong toxins that will cause diarrhea and vomiting and should not be consumed. European buckthorn fruit are extremely unpleasant-tasting, have a strong laxative effect, and are considered poisonous. Alder-leaved buckthorn, a native plant common throughout Canada and the U.S, was used medicinally as an infusion to relieve pain and swelling, purify the blood or treat constipation and gonorrhea.

EDIBILITY: edible with caution (toxic), poisonous

FRUIT: Berry-like drupes 6–9 mm long, often ripening unevenly in bunches, turning from green to yellow to a purple or bluish-black colour.

SEASON: Flowers June to July. Fruits ripen August to September.

DESCRIPTION: Erect or spreading shrub or small tree, 0.5–6 m tall, with oval to elliptic, prominently veined, 5–12 cm-long leaves. Flowers inconspicuous, greenish yellow, all male or all female in a cluster, forming flat-topped clusters in axils of lower leaves (away from the branch tips).

Alder-leaved buckthorn (*R. alnifolia*) is a medium-sized shrub native to Canada, 0.5–1.5 m tall, with clusters of 2–5 flowers. Stems are unarmed, green when young and maturing to purplish grey. Leaves oval, 2–10 cm long, 1–5 cm wide, short-stalked, margins toothed or wavy, prominently 10- to 14-veined. Grows in moist, open to shady meadows, swamps and on streambanks throughout Ontario. Also called: dwarf alder.

European buckthorn (*R. cathartica*) is generally a small tree, to 6 m tall. Branches grey and tipped with a small thorn. Leaves predominantly opposite, 1–4 cm wide, 3–8 cm long, oval to elliptic lateral veins in 3 pairs, margins smooth. Flowers borne in flat-topped clusters of 5–25 in leaf axils. Fruit a black, berry-like drupe. Found in moist or dry habitats along streambanks, roadsides and waste areas where it has escaped from cultivation and naturalized in southern Ontario.

Glossy buckthorn (*R. frangula*) is a shrub or small tree growing to 6 m with a rounded crown. Bark thin, brown to grey and covered in pale lenticels. Branches minutely hairy, brown to grey bark with lenticels, buds do not have scales. Leaves glossy green to 8 cm long and 5 cm wide with 5–10 parallel veins per side, smooth margins, on 2 cm stalks. Flowers single or in clusters of 2–8. Fruits slightly poisonous, to 7 mm across, 2–3 nut-like stones per fruit. Inhabits moist, shady sites around bogs and in woods and ravines around urban areas of southern and eastern Ontario. Also called: alder buckthorn, buckthorn • syn. *Frangula alnus*.

Alder-leaved buckthorn (*R. alnifolia*)

Glossy buckthorn (*R. frangula*)

185

Canadian Yew

Taxus canadensis

Also called: ground hemlock

Canadian yew (*T. canadensis*)

The showy berry-like fruit of this species, with its sweet taste but slimy texture, has historically been considered edible. However, the hard seeds found within the fleshy cup are extremely poisonous, so this fruit is not recommended for consumption (see Warning). The leaves are also toxic. The bark of western yew (native to BC and parts of Alberta) is the original source of the anticancer drug taxol. After a long period of development by the National Cancer Institute and pharmaceutical partners, taxol was approved for use in treating a variety of cancers and is particularly successful in treating breast and ovarian cancers that historically had extremely low survival rates. The slow-growing western yew, however, became quickly depleted in the wild by unregulated over-harvesting. *Taxane* derivatives are now in great part obtained from managed harvest of the more common Canada yew in eastern Canada, from which the drug is prepared by extraction and semi-synthesis.

Some native peoples used yew bark for treating illness (indeed, this is how modern researchers first knew to research this plant) and applied the wet needles as poultices on wounds, but do not try this remedy (see Warning). The Ojibwa treated rheumatism by boiling Canada yew and eastern red cedar twigs together and either drinking the resulting decoction or sprinkling it on hot rocks to produce steam. The Algonquin boiled the needles, sometimes with pin cherry, to treat rheumatism and to take as a tea after childbirth. The Montagnais used a tea of Canada yew to treat fevers and general weakness, and the Cree used the tea to treat stomachaches and menstrual cramps. The bark was used by the Mi'kmaq to treat bowel and internal troubles, and a tea of the needles for fever. With their dark, evergreen needles and scarlet berries, yews make lovely ornamental shrubs, but their poisonous seeds, branches and leaves are dangerous to children and some animals (see Warning).

EDIBILITY: poisonous

FRUIT: Berry-like arils (fleshy cones in leaf axils), 4–5 mm across, with a cup of orange to red fleshy tissue around the single bony seed, open at the top (looks like a berry with a hole on top)

SEASON: Flowers in May and June. Berries ripen to an orange or deep red from July to October.

DESCRIPTION: Evergreen shrub or small tree rarely growing over 2 m tall, often low and spreading. Branches initially green then becoming reddish brown and scaly as they mature. Linear leaves to 25 mm long and 3 mm wide with a narrowed sharp tip, appearing to be arranged in 2 rows or spirally around stem. Male and female cones are usually on the same shrub; the male cones inconspicuous, the female fruits scarlet arils about 7.5 mm long. Grows in dry to moist, rich forests and wetlands (swamps, bogs) at low elevation.

Canadian yew (*T. canadensis*)

WARNING: *The needles, bark and seeds contain extremely poisonous, heart-depressing alkaloids called* taxanes. *Drinking yew tea or eating as few as 50 leaves can cause death. The berries (which take 2 years to mature) are eaten by many birds, and the branches are said to be a preferred winter browse for moose, but many horses, cattle, sheep, goats, pigs and deer have been poisoned from eating yew shrubs, especially when the branches were previously cut.*

Western poison-ivy (*T. rydbergii*).

The scientific name for this genus comes from the Latin *toxicum*, meaning "poison," and the Greek *dendron*, meaning "tree." Poison-ivy and poison-oak plants contain an oily resin (*urushiol*) that causes a nasty skin reaction in most people, especially on sensitive skin and mucous membranes. Since the allergic contact *dermatitis* appears with some delay after exposure, many people do not realize that they have come into contact with these plants until it is too late. Sensitization can also lead to a more severe reaction after repeated exposure. *Urushiol* is not volatile and therefore it is not transmitted through the air, but it can be carried to unsuspecting victims by pets or through clothing and tools and even on smoke particles from burning poison-ivy or poison-oak plants. The resin can persist on pets and clothing for months and is also ejected in fine droplets into the air when the plants are pulled. Washing with a strong soap can prevent a reaction if it is done shortly after contact since the resin can be removed by this process. Washing

Western poison-ivy (*T. rydbergii*).

also prevents transfer of the resin to other parts of the body or to other people. Be sure to use cold water as warm water can help the resin to penetrate into your skin where it is extremely difficult to remove. Do not burn these plants to get rid of them as the smoke contains the allergic compounds and can be very dangerous if inhaled. The liquid that oozes from poison-ivy or poison-oak blisters on affected skin does not contain the allergen. Ointments and even household ammonia can be used to relieve the itching of mild cases, but people with severe reactions might need to consult a doctor. If you see "leaves of three, let it be!"

EDIBILITY: poisonous

FRUIT: Fruits dry, whitish, ridged berry-like drupes 4–5 mm wide, in upright clusters.

SEASON: Flowers May to July. Fruits ripen August to September.

DESCRIPTION: Trailing to erect deciduous shrubs, forming colonies. Leaves alternate, bright glossy green, divided into 3 oval leaflets, scarlet in autumn. Flowers cream-coloured, 5-petalled, 1–3 mm across, forming clusters.

Eastern poison-ivy (*T. radicans*) grows to less than 1 m tall and has well-developed stems and so often forms thickets. Leaflets are irregularly toothed or slightly lobed, to 15 cm long and 6 cm wide. Inhabits moist, sandy or rocky soil, in dry, open woods in partial sun in southern and occasionally western and central Ontario. Also called: *Rhus radicans*.

Vining poison-ivy (*T. radicans* var. *radicans*) is virtually identical to Eastern poison ivy except that it has a climbing, vining habit. Stems thick and reddish. Found in southern Ontario. Also called: syn. *T. radicans* ssp. *negundo*.

Western poison ivy (*T. rydbergii*) is a trailing to erect shrub growing to 2 m tall and forming patches or thickets. Leaves bright glossy green, resinous, entire (neither toothed nor lobed), scarlet in autumn. Grows on dry rocky slopes of southern Ontario. Also called: northern poison oak • syn. *T. radicans* var. *rydbergii*, *Rhus radicans* var. *rydbergii*.

Eastern poison-ivy (*T. radicans*)

Vining poison-ivy (*T. radicans* var. *radicans*)

189

Poison Sumac *Toxicodendron vernix*

Also called: *Rhus vernix*

Poison sumac (*T. vernix*)

Unwary harvesters gathering the beautiful white fruits or arching leaves for floral arrangements have been nastily surprised by this toxic plant. Most people develop a severe skin reaction similar to that produced by poison-ivy after coming into contact with this species. Recognized as poisonous by indigenous peoples of North America, few groups used it medicinally and most avoided it. See Toxicodendron spp. for more information.

Poison sumac has smooth-edged leaves and white (not orange or red) fruit, which are important identifica-tion traits that distinguish it from the similar-looking but non-toxic true sumacs, prickly ashes and elderberries.

EDIBILITY: poisonous

FRUIT: Glossy, white drupes 4–6 mm in diameter, dry (not juicy) in arching or hanging clusters.

SEASON: Flowers June to July. Fruits ripen July to August, often persisting through winter.

DESCRIPTION: Small tree or shrub to 6 m tall, heavily branched but less so near the base. Bark greyish brown, blotchy, with raised bumps. Branches dotted with lenticels, often emerging from base of main trunk. Stems red, emitting a dark toxic sap if broken or injured. Leaves smooth-edged, alternate, compound, 7–13 leaflets, on a central stalk 15–30 cm long. Flowers tiny, male and female on separate plants, on elongated clusters to 20 cm long, yellowish-green. Inhabits shady damp areas and open boggy wood-lands, including swamp edges and poorly drained bottomlands and woods in southern Ontario.

Poison sumac (*T. vernix*)

Poison sumac (*T. vernix*)

191

Devil's Club   *Oplopanax horridus*

Also called: *Echinopanax horridum*

Devil's club (*O. horridus*)

There are widely differing reports as to the both the edibility of these berries and their medicinal properties, so these fruit are not recommended for consumption. Throughout its range, Devil's club (which is botanically-related to the ginseng family) is considered to be one of the most powerful and important of all medicinal plants. First Nations people considered the berries of this plant to be inedible, perhaps partly because they are held aloft above a remarkable fortress of irritating spiny leaves and stems and because even the berries have spikes!

Although Devil's club tea is recommended today for binge-eaters who are trying to lose weight, some tribes used it to improve appetite and to help people gain weight. In fact, it was said that a patient could gain too much weight if it was used for too long. Some tribes used a strong decoction of the plant to induce vomiting in purifying rituals preceding important events such as hunting or war expeditions. This decoction was also applied to wounds to combat staphylococcus infections, and ashes from burned stems were sometimes mixed with grease to make salves to heal swellings and weeping sores.

Like many members of the ginseng family, Devil's club contains glycosides that are said to reduce metabolic stress and thus improve one's sense of well-being. The roots and bark of this plant contain the majority of active compounds, and they have traditionally been used to treat arthritis, diabetes, rheumatism, digestive troubles, gonorrhea and ulcers. The root tea has been reported to stimulate the respiratory tract and to help bring up phlegm when treating colds, bronchitis and pneumonia. It has been used to treat diabetes because it helps regulate blood sugar levels and reduce the craving for sugar. Indeed, Devil's club extracts have successfully lowered blood sugar levels in laboratory animals.

EDIBILITY: poisonous, and not recommended

FRUIT: Fruits bright red, berry-like drupes, slightly flattened, sometimes spiny, longer (5–8 mm) than wide, in showy pyramidal terminal clusters.

SEASON: Flowers May to July. Fruits ripen July to September.

Devil's club (*O. horridus*)

DESCRIPTION: Strong-smelling, deciduous shrub, 1–3 m tall, with spiny, erect or sprawling stems. Leaves broadly maple-like, 10–40 cm wide, with prickly ribs and long, bristly stalks. Spines on leaves and stems grow up to 1 cm long. Flowers greenish white, 5–6 mm long, 5-petalled, forming 10–25 cm-long, pyramid-shaped clusters of bright red berries. Grows in moist, shady foothill and sun-dappled forest sites along the northern shore of Lake Superior. Easy to find when you lose your footing on a hilly trail as it is invariably the plant that you grab onto to stop your fall.

Devil's club (*O. horridus*)

WARNING: *Devil's club spines are brittle and break off easily, embedding in the skin and causing infection. Some people have an allergic reaction to the scratches from this plant. Wilted leaves can be toxic so only fresh or completely dried leaves should be used to make a medicinal tea, but even then the tea should be taken under the guidance of a registered herbalist and in moderation because extended use can irritate the stomach and bowels.*

Honeysuckles *Lonicera* spp.

Black twinberry (*L. involucrata*)

Most First Nations considered honeysuckle berries to be inedible or poisonous. The flowers, however, produce sweet nectar at their base that can be sucked out. This treat was especially enjoyed by children. Although some honeysuckle berries are nauseously bitter, others are mildly pleasant, though not tasty enough to go out of your way to find them. The striking blue berries of mountain fly honeysuckle are sweet-tasting with a bitter after taste, but they reportedly make a reasonable (and interestingly coloured!) jam and are also considered good in baked goods. The small size of honeysuckle berries, almost always less than 1 cm wide, is a disadvantage to using this fruit.

Honeysuckle was employed by a number of First Nations for a diverse range of ailments. An infusion of the bark of twining honeysuckle was used as a cathartic and diuretic to treat kidney stones, menstrual difficulties and dysuria. A tea made from its peeled, internodal stems was used for urine retention, flu and blood clotting after childbirth. A decoction of the root was mixed with other ingredients to treat gonorrhea. A tea from the berries and bark was ingested by pregnant women to expel worms. A decoction of the whole plant was given to children for fevers and general sickness. Black twinberry bark was taken for coughs and its leaves chewed and applied externally to itchy skin, boils and gonorrhoeal sores. Berry tea was used as a cathartic and emetic as a means to purify the body and cleanse the stomach and chest. A decoction of twinberry leaves or inner bark was used as daily eyewash to bathe sore eyes. Boiled bark was applied to burns, infections and wounds. In spite of the myriad traditional medical uses of honeysuckle, however, the plant is rarely used today in modern herbalism.

The stems, which are supple and relatively strong, were used by First Nations as building materials and to make fibres for mats, baskets, bags, blankets and toys. Children used the hollow stems as straws. The plant was used in a number of love potions, charms and medicines to either form or destroy a relationship. During Victorian times, teenage girls were told not to bring honeysuckle home in the belief that the flowers induced erotic dreams.

EDIBILITY: edible, not palatable, edible with caution (toxic) or poisonous

Black twinberry (*L. involucrata*)

FRUIT: Fruit an orange, red, yellow, blue or black berry to 1 cm across containing several seeds.

SEASON: Flowers May to July. Fruits ripen July to September.

DESCRIPTION: Twining, deciduous woody vines or shrubs with trailing or climbing habits or growing to 5 m or more in height, with opposite, simple, smooth-margined leaves that are more or less oval, 3–15 cm long. Flowers sweet-scented, bell-shaped or tubular, 1–3 cm long, with a sweet, edible nectar.

Black twinberry (*L. involucrata*) grows to 5 m tall but usually shorter, with 4-angled twigs that are greenish when young, greyish with shredding bark when older. Leaves oval, to 16 cm long and 8 cm wide, sharp-pointed at tip. Flowers bell-shaped, yellow, 1–2 cm long, in pairs surrounded by fused, green to purple bracts (an "involucre," hence the scientific name involucrata). Fruits shiny, black berries, to 1.2 cm across. Berries are bitter-tasting and reported as inedible to poisonous. Grows in moist or wet soil in forests, clearings, riverbanks, swamps and thickets of northern Ontario down to

Black twinberry (*L. involucrata*)

Lake Superior. Also called: bracted honeysuckle, northern honeysuckle, twinflower honeysuckle.

Fly honeysuckle (*L. canadensis*) is a straggly, somewhat erect shrub growing to 1.5 m tall with a similar width. Branches smooth and greenish/purplish when young, becoming grey/brown with shredding bark when mature. Leaves egg-shaped to elliptic with a blunt or pointed tip, 3–9 cm long, bright green above, paler below; edges smooth, fringed with fine hairs.

Mountain fly honeysuckle (*L. villosa*)

Fruit red, paired, egg-shaped, connected to a slender stalk growing from the leaf axil. Fruit grow sparsely on the bush and are considered mildly toxic by some sources. Inhabits forested areas and swampy thickets with moderate sunlight. Particularly abundant around the Great Lakes, it can be found across the province but rarely north of the 50th parallel. Also called: American or Canadian fly honeysuckle.

Hairy honeysuckle (*L. hirsuta*) is a vine growing to 3 m. Branchlets greenish to purplish with purplish-brown spots and glandular-tipped hairs. Mature branches grey or brown with shredding bark. Young stems hairy with no waxy bloom. Leaves widely oval, pointed to blunt tips, round and tapered at base, 5–13 cm long and 2.5–9 cm wide, rough, downy hair on leaf surface and margins. Upper leaf pairs saucer-like, fused. Flowers 15–25 mm long, yellow to orange, turning reddish when older, growing in whorled clusters on a short stem above leaves. Fruits orange-red, clustered on short stems at the centre of leaves at stem tip, many-seeded. Inhabits dry to moist sandy or clay soils in swampy areas and forests. Less abundant than fly honeysuckle but with a similar range.

Mountain fly honeysuckle (*L. villosa*) is a small shrub to 1 m high with an open growth habit similar to that of an azalea. Branches purplish to green, finely hairy when young and maturing to reddish-brown or grey with shredding bark. Leaves hairy on both surfaces, edges hairy, green above and lighter beneath, about 5 cm long and 2.5 cm wide, growing closely bunched together at stem tips. Flowers pale yellow. Fruit in pairs (not fused), egg-shaped, striking deep blue colour with a whitish bloom. Berries are considered edible, but have a bitter aftertaste. Found throughout the province inhabiting moist areas such as lakeshores, swamp edges and peat bogs as well as rocky (but moist) places and sun-dappled mixedwood and coniferous forests.

Swamp fly honeysuckle (*L. oblongifolia*) is a small shrub resembling fly honeysuckle, growing to 1.5 m or less. Branches purplish/greenish, smooth to lightly hairy when young, maturing to grey with shredding bark. Leaves narrow, opposite, to 6 cm long, 2.5 cm wide, on short stemlets (*petioles*), finely downy when young, hairless when mature. Flowers pale yellow, tubular (12 mm long) with two lips, growing on a short stalk from leaf axils. Fruit round, translucent, red to orange or purplish-red, paired (but often of differing sizes), sometimes joined at the base. Fruit is considered inedible. Inhabits rocky, moist, cool coniferous woods, marsh edges and streambanks. Common in southern Ontario through Georgian Bay to Lake Superior and rare in the boreal forest.

Tartarian honeysuckle (*L. tatarica*) is the most common of several ornamental honeysuckles now growing wild in Ontario. It is an erect shrub to 3 m high with thin green leaves, smooth on both sides with a blunt or pointed tip and rounded or slightly heart-shaped bases, 2–6 cm long. Flowers in pairs, separate or fused, pink or white with a short tube and showy lobes. Fruit pairs partly united at the base, berries either red, orange or yellow. Found in fields, open woods and wastelands of southern Ontario.

Twining honeysuckle (*L. dioica*) is a twining, woody vine growing to 3 m. Young stems green to purplish-red with a waxy bloom, older stems grey or brown with shredding bark. Leaves not hairy on the margins, and the upper pairs are fused at the base to appear disc- or saucer-like. Flowers yellow to orange (sometimes dark reddish with age), growing clustered on a short stem from the centre of a terminal cup. Berries smooth, oval, reddish, growing in a cluster from the terminal leaf "disc." Fruit are extremely bitter and not recommended for eating. Grows in dry woods, thickets and rocky slopes of Ontario, most abundant in the south. Also called: limber honeysuckle, red honeysuckle, glaucous honeysuckle.

Twining honeysuckle (*L. dioica*)

Twining honeysuckle (*L. dioica*)

Snowberries   *Symphoricarpos* spp.

See also: *Gaultheria hispidula*—creeping snowberry

Common snowberry (*S. albus*)

Some sources report that these berries are edible, though not very good. However, snowberries are toxic and in large quantities can be mildly poisonous. Though a few Native North American groups consumed the fruits in times of scarcity, they are widely considered poisonous and were more commonly used medicinally, especially common snowberry, which served to treat sores, rashes, burns and infections.

Despite their toxicity, the spongy white berries are fun to squish and pop—rather like bubble wrap! The unusual fruits persist on the plant through winter, providing a showy and decorative display that in mild winters can last well into spring. The berries make a wonderful addition to winter holiday wreaths, garlands and other festive decorations.

This is a very drought-tolerant and decorative species that will thrive on steep slopes and other areas that may otherwise be difficult to landscape. The leaves and flowers, albeit small, are pretty and the white berries provide winter forage for birds and small mammals while giving a showy winter display.

EDIBILITY: edible with caution (toxic), poisonous

FRUIT: White, waxy, spongy berry-like drupes, 6–10 mm long, singly or in clusters on stem tips.

SEASON: Flowers May to August. Berries ripen and whiten August and September.

DESCRIPTION: Erect, deciduous shrubs usually 0.5–1 m tall. Leaves simple, opposite, elliptic to oval, 2–8 cm long and 1–5 cm wide. Stems flexible and strong, grey in colour, with bark becoming shredded on more mature specimens. Flowers pretty pink to white, broadly funnel-shaped, 4–7 mm long, borne in small clusters at the stem tips. Spreads rapidly through a tough, dense, underground

Common snowberry (*S. albus*)

Common snowberry (*S. albus*)

root system (rhizomes) and quickly forms an impenetrable thicket if left alone. Inhabits rocky banks, hedgerows, forest edges and roadsides.

Northern snowberry (*S. occidentalis*) is a small shrub to 1 m high, spreading by stolons to form dense thickets. Leaves larger than common snowberry, to 8 cm long, with edges either shallowly lobed, wavy-toothed or smooth. Flowers pale pink, about 9 mm long, borne in clusters of 2–10 in the leaf axils. Young drupes greenish-white, becoming blackish in August and September. Grows in rocky and sandy fields and clearings as well as disturbed sites, perhaps introduced from the west since it's found in dispersed localities. Also called: wolfberry, buckbrush.

Common snowberry (*S. albus*) grows up to 1 m tall, forming low thickets. Stems flexible and strong, light brown at first then turning purple-grey with bark becoming shredded on more mature specimens. Dark green leaves to 5 cm long with smooth edges. Flowers pretty pink to white, broadly funnel-shaped, 4–7 mm long, borne in small clusters at the stem tips. Spreads rapidly through a tough, dense, underground root system (rhizomes) and quickly forms an impenetrable thicket if left alone. Inhabits rocky banks, hedgerows, forest edges and roadsides from Ottawa to Thunder Bay.

Northern snowberry (*S. occidentalis*)

WARNING & WILD GARDENING: *The branches, leaves and roots of snowberries are poisonous, containing the alkaloid chelidonine, which can cause vomiting, diarrhea, depression and sedation. Most tribes considered snowberries poisonous and did not eat them. Because of its drought tolerance, tenacious roots and thick growth habit, snowberry is an ideal planting to stabilize slopes and provide vegetation in difficult-to-plant areas.*

Northern snowberry (*S. occidentalis*)

Common snowberry (*S. albus*)

Northern snowberry (*S. occidentalis*)

Baneberries *Actaea* spp.

Red baneberry (*A. rubra*)

Baneberries are related to the commercial phytomedicine black cohosh, and some indigenous peoples used baneberry root tea in a similar way to treat menstrual and postpartum problems, as well as colds, coughs, rheumatism and syphilis. Herbalists have used baneberry roots as a strong antispasmodic, anti-inflammatory, vasodilator and sedative, usually for treating menstrual cramps and menopausal discomforts. The berries, like other parts of the plant, are poisonous and inedible (see Warning).

Baneberry is a striking-looking plant with its attractive foliage and delicate stems of puffy white flowers in spring followed by showy spikes of red or white berries in fall. Planted with ferns, hostas and other shade-loving species, it makes for a decorative addition to the shade garden.

The two species native to Ontario, white and red baneberry, are closely related and look very similar. The fact that white varieties of red baneberry occur over the same range further complicates differentiating the two. If you're not sure, look at the berry stalks, which are red and thick on white baneberry, or black and thin on red baneberry.

EDIBILITY: poisonous

Red baneberry (*A. rubra*)

FRUIT: Fruits are very showy, several-seeded, glossy red or white berries 6–11 mm long, with a black dot at the tip, growing singly on a stalk in a cluster at the end of flowering stems.

SEASON: Flowers May to July. Ripens July to August.

DESCRIPTION: Branched, leafy, generally solitary perennial herbs, 30 cm–1 m tall, from a woody stem-base and fibrous roots. Stems long, wiry. Leaves alternate, few and large, crowded at base of stem and sparser near the top, each divided 2–3 times in three leaflets. Flowers white, with 5–15 sepals and petals, forming long-stalked, rounded clusters.

Red baneberry (*A. rubra*) is very similar to white baneberry except that the berry-bearing stalks are slender and black, and the black dot is smaller (less than 15 mm across). Also, the leaflets are usually shorter (about 6 cm), and the berries are usually red but sometimes white. Though these plants are uncommon in the south, they occur more frequently where red baneberry's range extends north into the boreal forest. Also called: snake berry, western baneberry • *A. arguta*, *A. eburnean*.

White baneberry (*A. pachypoda*)

White baneberry (*A. pachypoda*)

White baneberry (*A. pachypoda*) has compound leaves with egg-shaped leaflets up to 10 cm long. Small flowers of 7–15 white petals with conspicuous white stamens. Smooth white berries with a black dot of 1.5–1.5 mm in diameter, borne on thick red stalks clustered at the end of flowering stems. Inhabits moist deciduous and mixed forests across southern parts of the province. Also called: Doll's eyes • *A. alba*.

WARNING: *All parts of baneberry are poisonous, but the roots and berries are most toxic. Indeed, the common name "baneberry" derives from the Anglo-Saxon "bana," which means "murderous." Eating as few as 2–6 berries can cause cramps and burning in the stomach, vomiting, bloody diarrhea, increased pulse, headaches and/or dizziness. Severe poisoning results in convulsions, paralysis of the respiratory system and cardiac arrest. No deaths have been reported in North America, probably because the berries are extremely bitter and unpleasant to eat.*

Nightshades *Solanum* spp.

Bittersweet nightshade (*S. dulcumara*)

While members of this genus, like Belladonna (or deadly nightshade), are infamous worldwide for their toxicity, many—including tomato, potato and eggplant—are important food plants, with others used for medicinal purposes. Two species of nightshades grow wild in Ontario, one native and one introduced, and their berries considered poisonous by both First Nations peoples and early settlers. The leaves and unripe, green or yellow berries of bittersweet nightshade contain toxic alkaloids that can irritate and damage the gastrointestinal system and liver as well as cause weakness, vomiting, convulsions, paralysis and even death. The ripened berries contain fewer alkaloids and can be eaten in small quantities but the taste isn't really worth the risk. The same is true for Eastern black nightshade except that the ripe (black) berries are edible when eaten raw in moderation or when used in pies or jams.

Medicinally, the roots of bittersweet nightshade were used by First Nations to treat nausea and stomach problems. In Europe, the stem extract has been used to treat a variety of conditions, including dermatitis, pain, asthma and cancer, and is approved in Germany for the topical treatment of chronic eczema. The young stems and leaves of nightshade are also prepared as a homeopathic remedy for backaches and joint pains, cough, diarrhea and eye inflammations.

EDIBILITY: poisonous when immature, edible in moderation when ripe.

FRUIT: Fruits are round or oblong, several-seeded, tomato-like berries that are green when immature and ripening to shiny red or black.

SEASON: Flowers June to September. Fruits ripen August to October.

DESCRIPTION: Annual or perennial herbs or vines 0.2–3 m in height, erect or climbing, with leafy stems that may become woody at the base. Leaves alternate, simple and ovate to lance-like, sometimes with two wing-like lobes at the base. Flowers of 5 white or purple-blue petals curved back with a yellow cone projecting from the centre.

Bittersweet nightshade (*S. dulcumara*) is a perennial vine with a woody base and slender stems climbing or trailing over other vegetation. Leaves oval to slightly heart-shaped, 2–10 cm in length, often with two deeply cut wing-like lobes near the base and smooth margins. Flowers star-like with blue to purple petals, 10–15 mm across, borne in clusters. Berries round to oblong, initially green and ripening to yellow then bright red, often remaining on the plant after the leaves have fallen. Introduced from Europe, it is common in thickets, clearings, fields and forest edges near habitation throughout southern Ontario. Also called: European bittersweet, bitter nightshade, climbing nightshade.

Eastern black nightshade (*S. ptychanthum*) is an erect annual or perennial herb 20–70 cm high with a taproot. Leaves lance-like, 5–10 cm long, dark green with smooth or bluntly toothed margins. Flowers usually less than 1 cm across with white petals. Green berries ripening to shiny black, up to 11 mm in diameter and arranged in loose clusters. Inhabits disturbed sites, clearings and cultivated fields of southern Ontario. Also called: West Indian nightshade.

Eastern black nightshade (*S. ptychanthum*)

Bittersweet nightshade (*S. dulcumara*)

Jack-in-the-pulpit (*A. triphyllum*)

The roots of this plant are reportedly nutritious but must be pounded and then dried to eliminate the toxic calcium oxalate contained in them before they can be eaten. The Iroquois used an infusion of the roots to induce temporary sterility in women and also used a compound snuff to treat headaches. A decoction or infusion of the roots was taken for aches and pains, and was used to treat diarrhea in children. A poultice of the plant was applied hot on face sores and bruises and to treat lameness. In combination with black snakeroot and cherry bark, this plant was used to treat coughs and fevers. An infusion of the root was used as a gargle to treat sore throats but was considered poisonous if swallowed.

Jack-in-the-pulpit is an extremely decorative plant with high architectural foliage, and the large clusters of bright red berries on a central stalk last well into fall. Birds such as ring-necked pheasant, wood thrush and wild turkey enjoy these berries.

EDIBILITY: mildly toxic, poisonous

FRUIT: An elongated cluster of tightly joined shiny orange to bright red berries, 1 cm in diameter each, 1–3 seeded.

SEASON: Flowers late May to June. Fruits ripen August to September.

DESCRIPTION: Erect perennial 10–80 cm tall with several papery sheaths at the base, growing from a thick, short corm. Leaf stalks long, usually 2 per plant, purplish-green, topped with 3 compound leaflets, with base sheathing the flowering stem. Leaflets broadly oval, pointed tip, to 18 cm long, joined directly to leaf stalk. Leaves normally died-back before fruits ripen. "Flower" is a distinctive purple-, brown- or white-striped spathe (green sheath) to 18 cm long, curling over and surrounding a central spadix (club-like structure bearing the actual tiny yellow flowers at its tip) usually 6–9 cm tall. Grows in moist areas in sun-dappled, deciduous forests, often with eastern hemlock and yellow birch, in southern and western Ontario.

WARNING: *All parts of this plant contain crystals of calcium oxalate, which cause intense burning in the mouth, throat, stomach and intestines if ingested.*

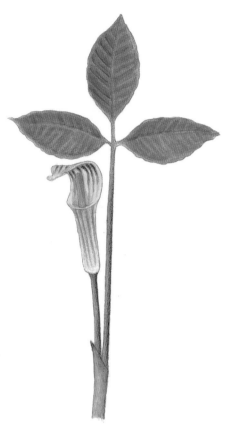

Jack-in-the-pulpit (*A. triphyllum*)

Jack-in-the-pulpit (*A. triphyllum*)

207

Pokeweed *Phytolacca americana*

Also called: American pokeweed, pokeberry, pokebush

Pokeweed (*P. americana*)

Pokeweed berries are considered poisonous and were rarely eaten by First Nations peoples. According to some accounts, the toxins are restricted to the seed or inactivated by cooking, but the cooked fruit should nonetheless be treated with caution. The leaves are also toxic but, if boiled twice in fresh water to remove toxins, are edible and quite tasty (see Warning). Apparently less toxic, the spring shoots were gathered and eaten as fresh greens by the Iroquois and Maliseet, but the shoots can also be blanched before eating or storing.

With a long history of medicinal use by First Nations and herbalists alike, the plant has been used as an emetic, laxative, expectorant and treatment for chest colds or bewitchment. Known for its immune-stimulating, antiviral, anti-inflammatory and antibiotic properties, pokeweed root preparations were applied to bruises, bumps and sprains while berry and shoot preparations were used to treat skin lumps and rheumatism or wounds, respectively. The toxic (and medicinal) chemicals are phytolaccatoxin, phytolaccin, triterpene saponins, histamines and oxalic acid. Producing multiple antiviral proteins, pokeweed is currently under study as a potential treatment for HIV/AIDS.

The term "poke" may be derived from the Algonquin, who applied a similar term to plants that contain dye. Appropriately, the ripe berries were also used to colour foods and beverages.

EDIBILITY: mildly toxic, poisonous

FRUIT: Round berry-like pomes with flat indented tops and bottoms, 6-11 mm in diameter, green when immature and ripening to white then a shiny purple-black, arranged in elongated clusters on pink stalks.

SEASON: Flowers June to September. Fruits ripen August to October.

DESCRIPTION: Perennial shrub to 3 m in height but often shorter, with a large white taproot and green stems. Leaves simple and large, oval to lance-like up to 30 cm long and producing an unpleasant odour when bruised or crushed. Flowers generally with 5 white or occasionally pink or purple petals, 2–5 mm across, on pink stalks loosely arranged around a long pink flowering stem (5–30 cm). Found in moist clearings, pastures, thickets and along roadsides across southern and eastern Ontario.

WARNING: *The leaves, berries, seeds and roots are said to be poisonous and can cause burning sensations in the mouth and throat, sweating, vomiting, diarrhea, blurred vision, unconsciousness and, rarely, death. Pokeweed contains several potentially toxic compounds, including mitogens, that stimulate the immune system and disrupt the body's white blood cells. Exposure to juices from this plant, either consumed or through open cuts or scrapes, can affect the blood and damage chromosomes. Use of this plant is not recommended.*

Pokeweed (*P. americana*)

Blue Cohoshes *Caulophyllum* spp.

Giant blue cohosh (*C. giganteum*)

Historically, the roasted seeds of blue cohoshes have been used as a caffeine-free coffee substitute. The attractive, berry-like fruits remain on the plant until autumn, but they should not be eaten, especially by children, because they are potentially poisonous. Even handling the roots, leaves or seeds can cause dermatitis in sensitive individuals.

Blue cohosh is one of the most important indigenous eastern North American plant medicines. The plant was used extensively by First Nations to facilitate labour, treat irregular menstruation, stimulate the uterus and, with correct dose and timing, induce abortions. It has been labelled "a powerful women's ally" by modern herbalists because of its historical use to treat various gynecological condi-

tions. It is used alone or in combination with other herbs to treat endometriosis, chlamydia and cervical dysplasia. As a homeopathic uterine tonic, it is mixed in a 1:1 ratio with black cohosh (*Cimicifuga racemosa*) and used as a tincture in the last 2–4 weeks of pregnancy to tone the uterus to facilitate childbirth, jumpstart a stalled labour and ease labour pains. If used during early pregnancy, the induced uterine contractions can cause a miscarriage or early delivery. The root contains a number of alkaloids and glycosides, of which the alkaloid methylcytisine and the glycoside caulosaponin seem to be the responsible bioactive constituents. Caulosaponin exerts its oxytocic (childbirth hastening) effects by causing muscle spasms resulting in contraction of the uterus. It also constricts heart blood vessels and has demonstrated cardiac muscle toxicity in animals. For this reason, blue cohosh should not be taken by people with hypertension and heart disease. Methylcytisine displays activity similar to nicotine in animals in that it elevates blood pressure and stimulates respiration and intestinal motility, but it is much less toxic. Other ailments for which blue cohosh root was prescribed include rheumatism, pelvic inflammatory disease, gout, dropsy, colic, sore throat, abdominal cramps, hiccoughs, epilepsy, hysterics, inflammation, urinary tract infections, lung ailments and fevers. It has been used as a diuretic and to expel intestinal worms.

The bright blue seeds have been used to make seed jewellery. These plants are occasionally grown as woodland garden ornamentals.

EDIBILITY: poisonous

FRUIT: Deep blue, berry-like fruit containing 2 seeds, on individual thick stalks, in loose terminal clusters.

SEASON: Flowers April to June. Fruits ripen July to August.

DESCRIPTION: Perennial, deciduous herbs from rhizomes, 30–90 cm tall, with large, compound leaves of stalked three leaflets, each lobed and ending in 3–5 distinct tips, on a single stem; larger stems have 2 leaves. Flowers borne on spikes, each having 6 greenish-yellow to purple, petal-like sepals, with 6 fleshy nectar glands at the base of each sepal.

Blue cohosh (*C. thalictroides*) has 5–70 inconspicuous, purplish-brown to yellow-green flowers, 1–2 cm wide, borne on a loosely branched cluster. Grows in low elevation, moist woods in southern Ontario. Also called: papoose root, squaw root, yellow-flowered blue cohosh.

Giant blue cohosh (*C. giganteum*) is considered by some botanists as a subspecies of blue cohosh. One of Ontario's earliest blooming wildflowers, this plant has 4–18 most often purple but also red, brown or yellow flowers, 1–4 cm wide, in early spring before C. thalictroides flowers. Grows in low elevation, moist woods in southern Ontario. Also called: purple-flowered blue cohosh.

Blue cohosh (*C. thalictroides*)

Blue cohosh (*C. thalictroides*)

May-apple *Podophyllum peltatum*

Also called: Indian apple, wild mandrake

May-apple (*P. peltatum*)

Although the roots and leaves of this plant are highly toxic, ripe may-apples can have a sweet, strawberry-like flavour with sweet and acidic pulp surrounding the seed. The fruit can be eaten raw, made into jellies, marmalades and pies or used as a flavouring for drinks—it's particularly tasty added to lemonade! First Nations peoples also dried the fruits into cakes for future consumption. However, wait for the fruit to ripen fully, right about when the plant begins to wither and die off. Otherwise, you'll likely encounter the immature fruit's laxative effects. Eating too much of the fruit can cause colic, and make sure to remove the rind, which, like other parts of the plant, are strongly cathartic.

Beyond a source of food, may-apple was widely used for medicinal purposes across its range and is one of North America's more powerful medicinal plants. Used for a broad range of symptoms and diseases, the plant's effects on the liver and digestive system are most notable. First Nations and early settlers considered the plant to be a "liver cleanser" and used it, particularly the root, to treat hepatits and jaundice, as a purgative and laxative and to treat fevers and syphilis. It is also used in homeopathy and herbology for gall bladder and intestinal problems.

The roots produce a resin called podophyllin that contains podophyl-lotoxin, a potent toxin that was developed into a leading anti-cancer agent called etoposide and included as an active agent in commerical treat-ment for genital and uterine warts. While podophyllotoxin and related chemicals can be helpful in treating testicular and small-cell lung cancers, it also kills healthy cells and is a teratogen (a substance that harms developing embryos). As such, may-apple should not be used during pregnancy or without qualified, professional supervision.

EDIBILITY: edible

FRUIT: Large yellow, lemon-shaped berry, 3–6 cm long, with several big, light brown seeds.

SEASON: Flowers April to June. Fruits ripen July to September.

DESCRIPTION: Perrenial, deciduous, herb to 60 cm tall from rhizomes to form dense stands. Stems with two opposite leaves, each large and green, radially lobed and up to 30 cm across. Fragrant flowers are solitary, nodding and borne between the two leaves; 5 cm in diameter with 6–9 white, waxy petals. Found in moist mixed and deciduous forests and shady meadows of southeastern Ontario.

WARNING: *All parts of this plant except the rindless fruits are toxic. Eating even tiny amounts of the roots or leaves is poisonous.*

May-apple (*P. peltatum*)

May-apple (*P. peltatum*)

Leatherwood *Dirca palustris*

Also called: moosewood, eastern leatherwood, wicopy

Leatherwood (*D. palustris*)

Because of its toxicity, leatherwood should not be eaten, but First Nations used the roots, bark and twigs for medicinal and practical purposes. An infusion of the inner bark of this species was taken as a laxative, and an infusion of the roots was used to treat pulmonary troubles. The Chippewa used a compound decoction of the root as a wash to strengthen hair and make it grow. The Iroquois used a compound infusion of the bark and roots as an analgesic for back pain and a decoction of the root and bark as a purgative. They also applied a decoc-tion of the branches as a poultice to reduce swelling of the limbs.

The Iroquois and other First Nations within the plant's range twisted the bark into cordage and used it for tying and binding. The common name "leatherwood" derives from the bark (not the wood), which is pliable yet strong.

EDIBILITY: poisonous

FRUIT: Purplish-red berry-like drupes round to elliptic, 9–12 mm long containing 1 brown pit.

Leatherwood (*D. palustris*)

SEASON: Flowers April to May. Fruits ripen June to July, falling soon afterwards.

DESCRIPTION: Deciduous, freely branching shrub growing to 2 m tall. Branchlets jointed, green when young, becoming brown to greyish-brown when mature. Bark tough and pliable. Leaves light green, smooth, 5–10 cm long, 3–7 cm wide, alternate, simple, margins toothless, egg-shaped blades broadly oval to elliptic, on a stalk less than 3 mm long. Flowers tubular, 6–9 mm long, pale yellow, in drooping clusters of 2–5. Inhabits rocky, loamy or sandy soils of moist or shaded woodlands and pastures of southern Ontario.

Leatherwood (*D. palustris*)

False Virginia Creeper *Parthenocissus vitacea*

Also called: grape woodbine, thicket creeper

False Virginia creeper (*P. vitacea*)

The berries (which contain high levels of oxalic acid) and foliage of this plant are considered poisonous. The Ojibwa, however, reportedly consumed the very closely related species, Virginia creeper (*Parthenocissus quinquefolia*), which also occurs in southern Ontario, most likely as escaped ornamentals. They boiled the stalks and ate the inner bark like corn on the cob. A syrup was also rendered by boiling the stalks and then used to cook wild rice.

The Iroquois used a decoction of false Virginia creeper to treat urinary problems and skin ailments. The most popular modern use for this plant is as an attractive garden climber or groundcover, particularly for its showy autumn foliage, which provides a vibrant display of fiery red, purple and scarlet leaves after the first frosts. It is also notable for its purplish-black berries and red stems, which remain on the plant after the leaves have fallen, providing an interesting winter display. False Virginia creeper is also useful as a wildlife attractant. The berries are a popular food for winter birds, in particular many species of songbirds, and wild deer and livestock browse its foliage. Because of its perennial root system and ground-covering habit, this plant has been used in slope stabilization, and it can provide valuable habitat for small mammals and birds.

EDIBILITY: poisonous

FRUIT: Fruits round, bright blue to purplish-black berries, 8–10 mm in diameter, with a thin layer of flesh around 3–4 seeds; when fruits ripen, stalks often bright red.

SEASON: Flowers June to July. Fruits ripen August to September.

DESCRIPTION: Woody, scrambling or climbing, deciduous vine reaching 20–30 m in length and height; tendrils sparsely branched. Leaves alternate, long-stalked (15–20 cm), palmately compound with 5 short-stalked leaflets; leaflets elliptic to oval, long-tapered at tips, wedge-shaped at base, 5–12 cm long, dark green and shiny above, paler beneath, short-stalked; margins coarsely and sharply toothed above middle; becoming brilliant scarlet in autumn. Flowers greenish, small, about 5 mm across;

COMPARE: *This plant is almost indistinguishable from true Virginia creeper (*V. quinquefolia*), the only major differences being that the tendrils of false Virginia creeper are less branched and lack the adhesive, sticky discs that allow the true Virginia creeper to climb smooth surfaces. As a result, false Virginia creeper rambles through trees and along the ground, attaching itself with twining tendrils, similar to those of a grape, rather than vertically adhering itself to walls, trellises and bare tree trunks. Rather than bunched in forked clusters, the berries of true Virginia creeper are arranged around a central axis, but the foliage (including the showy autumn colouring) of these two species is identical.*

25–200 or more in forked, branching clusters. Found in moist soils in woods and thickets and open ground along roadsides in southern Ontario.

False Virginia creeper (*P. vitacea*)

Canada moonseed (*M. canadense*)

The fruit of this species contains dauricine, an alkaloid that affects "excitable" muscles such as the heart and the smooth muscles of blood vessels. By blocking calcium flow, dauricine can decrease blood pressure and control the heart rate. A tincture of moonseed root was historically prescribed to treat scrofula, rheumatism, gout and cutaneous skin diseases. The plant was also used by Native Americans to treat sores or skin conditions, for feminine or stomach problems and as a laxative.

EDIBILITY: poisonous

FRUIT: Fruits dark blue, purple to black drupes, 6–7 mm, growing in loose grape-like clusters, containing a single crescent-shaped seed.

SEASON: Flowers June to July. Fruits ripen August to September.

DESCRIPTION: Perennial vine to

7 m, climbing trees or along the ground. Stems thin, twining, dark maroon, woody below. Young stems greenish or reddish, flexible, hairy. Climbs by twining around other plants and structures and has no tendrils. Leaves alternate, simple, 5–11 cm long and 5–15 cm broad with 3, 5 or 7 shallow lobes and heart-shaped bases, smooth margins, upper surface hairy, lower leaf surface silvery green and mildly hairy. Small creamy white flowers in loose clusters from leaf axils. Inhabits sunny areas in moist woods, streambanks, hedges and thickets in southern and eastern Ontario.

WARNING: *Do not confuse this species with wild grapes (*Vitis *species), a vine that is related but with only a superficially similar appearance. The dark blue, berry-like fruits of moonseed are highly poisonous and may be mistaken for edible grapes, particularly by children. The leaves of moonseed are smooth rather than toothed, and the seed is a single, crescent-shaped body in each fruit rather than the many seeds found in wild grapes.*

Canada moonseed (*M. canadense*)

Canada moonseed (*M. canadense*)

Glossary

accessory fruit: a fruit that develops from the thickened calyx of the flower rather than from the ovary (e.g., soapberry).

achene: a small, dry fruit that doesn't split open; often seed-like in appearance; distinguished from a nutlet by its relatively thin wall.

alkaloid: any of a group of bitter-tasting, usually mildly alkaline plant chemicals. Many alkaloids affect the nervous system.

alternate: situated singly at each node or joint (e.g., as leaves on a stem) or regularly between other organs (e.g., as stamens alternate with petals).

anaphylaxis: a hypersensitivity reaction to the ingestion or injection of a substance (a protein or drug) resulting from prior contact with a substance. Anaphylaxis can progress rapidly and be life-threatening.

annual: a plant that completes its life cycle in one growing season.

anthers: the pollen-producing sacs of stamens.

anthraquinone: an organic compound found in some plants that has a laxative effect when ingested. It is also used commercially as a dye and pigment, and also in the pulp and paper industry.

aril: a specialized cover attached to a mature seed.

armed: a plant furnished with defensive bristles or thorns.

Vascular Plant Parts

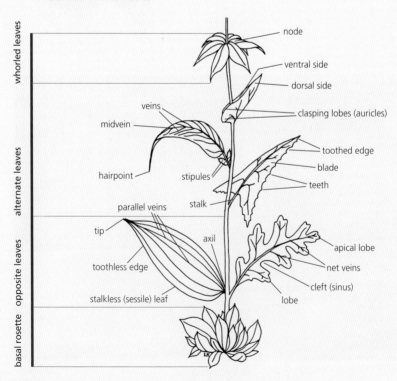

whorled leaves · alternate leaves · opposite leaves · basal rosette

node
ventral side
dorsal side
clasping lobes (auricles)
veins
midvein
toothed edge
blade
teeth
hairpoint
stipules
parallel veins
stalk
tip
axil
apical lobe
net veins
toothless edge
cleft (sinus)
stalkless (sessile) leaf
lobe

axil: the position between a side organ (e.g., a leaf) and the part to which it is attached (e.g., a stem).

berry: a fleshy, simple fruit that contains one or more ovule-bearing structures (carpels) that each contains one or more seeds; the outside covering (endocarp) of a berry is generally soft, moist and fleshy (e.g., blueberry).

biennial: a plant that lives for two years, usually producing flowers and seed in the second year.

bitters: alcoholic drinks consumed with a meal which contain bitter herbs to aid in the process of digestion (e.g., Swedish bitters or Angostura bitters).

bog: a peat-covered wetland characterized by *Sphagnum* mosses, heath shrubs and sometimes trees.

bract: a specialized leaf with a flower (or sometimes a flower cluster) arising from its axil.

calcareous: a type of soil with a high calcium content.

calyx: the outer (lowermost) circle of floral parts; composed of separate or fused lobes called sepals; usually green and leaf-like.

carpel: a fertile leaf bearing the undeveloped seed(s); one or more carpels join together to form a pistil.

cathartic: a substance that purges the bowels.

compound leaf: a leaf composed of two or more leaflets.

compound drupe: a collection of tiny fruit that forms within the same flower from individual ovaries; this type of fruit is often crunchy and seedy (e.g., boysenberries).

cone: a fruit that is made up of scales (sporophylls) that are arranged in a spiral or overlapping pattern around a central core, and in which the seeds develop between the scales (e.g., juniper).

corolla: the second circle of floral parts, composed of separate or fused lobes called petals; usually conspicuous in size and colour, but sometimes small or absent.

cultivar: a plant or animal originating in cultivation (e.g., loganberry or Golden Delicious apple).

deciduous: having structures (leaves, petals, seeds, etc.) that are shed at maturity and in autumn.

drupe: a fruit with an outer fleshy part covered by a thin skin and surrounding a hard or bony stone that encloses a single seed (e.g., a plum).

drupelet: a tiny drupe; part of an aggregate fruit such as a raspberry.

emetic: induces vomiting.

endocarp: the inner layer of the pericarp.

eulachon grease: grease from the eulachon (*Thaleichthys pacificus*), a small species of fish in the smelt family that lives most of its life in the Pacific Ocean but comes inland to fresh water for spawning.

fruit: a ripened ovary, together with any other structures that ripen with it as a unit.

glabrous: without hair, smooth.

glandular: associated with a gland (e.g., glandular hair).

glaucous: a frosted appearance due to a whitish powdery or waxy coating.

globose: shaped like a sphere.

glycoside: a two-parted molecule composed of a sugar and an aglycone, usually becoming poisonous when digested and the sugar is separated from its poisonous aglycone.

habitat: where a plant or animal is normally found; the characteristic environmental conditions in which a species is normally found.

haw: the fruit of a hawthorn, usually with a fleshy outer layer enclosing many dry seeds.

heath: a member of the heath family (Ericaceae).

herbaceous: a plant or plant part lacking lignified (woody) tissues.

hip: a fruit composed of a collection of bony seeds (achenes), each of which comes from a single pistil, covered by a fleshy receptacle that is contracted at the mouth (e.g., rose hip).

hybrid: a cross between two species.

hybridize: breeding together different species or varieties of plants or animals; the resulting hybrid often has characteristics of both parents.

inflorescence: flower cluster.

involucre: a set of bracts closely associated with one another, encircling and immediately below a flower cluster.

lanceolate: a long leaf that is widest at the middle and pointed at the tip.

lenticel: a slightly raised pore on root, trunk or branch bark.

mesic: habitat with intermediate moisture levels—not too dry or too moist.

montane: mountainous habitat, below the timberline.

multiple fruit: ripens from a number of separate flowers that grow closely together, each with its own pistil (e.g., mulberry, fig).

node: the place where a leaf or branch is attached.

nutlet: a small, hard, dry, one-seeded fruit or part of a fruit; does not split open.

opposite: situated across from each other at the same node (not alternate or whorled); or situated directly in front of another organ (e.g., stamens opposite petals).

ovary: the part of the pistil that contains the ovules.

ovules: the organs that develop into seeds after fertilization.

palmate: divided into three or more lobes or leaflets diverging from a common point, like fingers on a hand.

peduncle: a flower or fruit stem.

pemmican: a mixture of finely pounded dried meat, fat and sometimes dried fruit.

perennial: a plant that lives for three or more years, usually flowering and fruiting for several years.

Section of a regular flower with numerous carpels

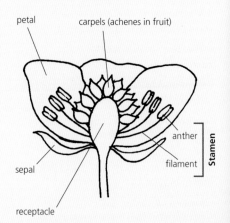

petal

carpels (achenes in fruit)

anther

Stamen

filament

sepal

receptacle

pericarp: the part of a fruit that derives from the ovary wall; generally consists of three layers: (from inside to outside) endocarp, mesocarp, exocarp.

petal: a unit of the corolla; usually brightly coloured to attract insects.

phytomedicine: the use of plants as medicine.

pinnate: with branches, lobes, leaflets or veins arranged on both sides of a central stalk or vein; feather-like.

pistil: the female part of the flower, composed of the stigma, style and ovary.

pitch: sticky tree sap, for example from a pine tree.

pome: a fleshy fruit with a core (e.g., an apple) comprised of an enlarged hypanthium around a compound ovary.

prostrate: growing flat along the ground.

purgative: causing watery evacuation of the bowels.

raceme: an unbranched cluster of stalked flowers on a common, elongated central stalk, blooming from the bottom up.

receptacle: an expanded stalk tip at the centre of a flower, bearing the floral organs or the small, crowded flowers of a head.

recurved: curved under (usually referring to leaf margins).

rhizome: an underground, often lengthened stem; distinguished from the root by the presence of nodes and buds or scale-like leaves.

saponin: any of a group of glycosides with steroid-like structure; found in many plants; causes diarrhea and vomiting when taken internally but commercially used in detergents.

sepal: one segment of the calyx; usually green and leaf-like.

spore: a reproductive body composed of one or several cells that is capable of asexual reproduction (doesn't require fertilization).

sporophyll: a spore-bearing leaf; a scale of a conifer cone.

spp.: abbreviation of "species" (plural).

spur: a hollow appendage on a petal or sepal, usually functioning as a nectary.

spur-shoot: a slow-growing, much-reduced shoot (e.g., on a larch or ginko tree).

stolon: a slender, prostrate, spreading branch, rooting and often developing new shoots and/or plants at its nodes or at the tip.

style: the part of the pistil connecting the stigma to the ovary; often elongated and stalk-like.

subalpine: just below the treeline, but above the foothills.

sucker: a shoot not originating from a seed, but from a rhizome or root.

tepal: a sepal or petal, when these structures are not easily distinguished.

throat: the opening into a corolla tube or calyx tube.

toxic: a substance that can cause damage, illness or death.

tundra: a habitat in which the subsoil remains frozen year-round characterized by low growth and lacking in trees.

unarmed: without prickles or thorns.

variety: a naturally occurring variant of a species; below the level of subspecies in biological classification.

References

Bennett, Jennifer, Ed. 1991. *Berries: A Harrowsmith Gardener's Guide*. Camden House Publishing, Camden East, Ontario.

Chambers, B., K. Legasy, & C. V. Bentley. 1996. *Forest Plants of Central Ontario*. Lone Pine Publishing, Edmonton, Alberta.

Dickinson, Timothy, Deborah Metsger, Jenny Bull & Richard Dickinson. 2004. *The ROM Field Guide to Wildflowers of Ontario*. Royal Ontario Museum é McClelland & Stewart, Toronto, Ontario.

Domico, Terry. 1979. *Wild Harvest: Edible Plants of the Pacific Northwest*. Hancock House Publishers, Saanichton, British Columbia.

Elias, Thomas S., & Peter A. Dykeman. 1990. *Edible Wild Plants: A North American Field Guide*. Sterling Publishing Company, New York, New York.

Hutchens, Alma R. 1991. *Indian Herbalogy of North America: The Definitive Guide to Native Medicinal Plants and Their Uses*. Shambhala Publications, Boston, Massachusetts.

Kershaw, Linda. 2001. *Trees of Ontario*. Lone Pine Publishing, Edmonton, Alberta.

Kuhnlein, Harriet V., & Nancy J. Turner. 1991. *Traditional Plant Foods of Canadian Indigenous Peoples: Nutrition, Botany and Use*. Gordon and Breach Science Publishers, Philadelphia, Pennsylvania.

Legasy, K., S. LaBelle-Beadman, & B. Chambers. 1995. *Forest Plants of Northeastern Ontario*. Lone Pine Publishing, Edmonton, Alberta.

Looman, J. & K.F. Best (1987) (rev.). *Budd's Flora of the Canadian Prairie Provinces*. Revised and enlarged edition from 1981 reprint. Agriculture Canada Publications, Hull, Québec.

MacKinnon, A., L. Kershaw, J. T. Arnason, P. Owen, A. Karst, & F. Hamersley Chambers. 2009. *Edible & Medicinal Plants of Canada*. Lone Pine Publishing, Edmonton, Alberta.

Marie-Victorin, Frère, Ernest Rouleau & Luc Brouillet. 2002. *Flore Laurentienne, 3ᵉ edition*. Presses de l'Université de Montréal, Montréal, Québec.

Marrone, Teresa. 2009. *Wild Berries & Fruits Field Guide: Minnesota, Wisconsin and Michigan*. Adventure Publications Inc., Cambridge, Minnesota.

Neal, Bill. 1992. *Gardener's Latin: A Lexicon*. Algonquin Books of Chapel Hill, Chapel Hill, North Carolina.

Newmaster, S. G., A. G. Harris, & L. J. Kershaw. 1997. *Wetland Plants of Ontario*. Lone Pine Publishing, Edmonton, Alberta.

Soper, James H. & Margaret L. Heimburger, 1981, *Shrubs of Ontario*. Royal Ontario Museum, Toronto, Ontario.

Stark, Raymond. 1981. *Guide to Indian Herbs*. Hancock House Publishers, North Vancouver, British Columbia.

Turner, N. J. 2005. *The Earth's Blanket: Traditional Teachings for Sustainable Living*. Douglas & McIntyre, Vancouver, British Columbia.

Turner, Nancy J., & Adam F. Szczawinski. 1988. *Edible Wild Fruits and Nuts of Canada*. Fitzhenry & Whiteside, Markham, Ontario.

Internet Sources

Canadian Biodiversity:
http://canadianbiodiversity.mcgill.ca/english/species/plants/index.htm

Database of Vascular Plants of Canada (VASCAN), from Canadensys:
http://data.canadensys.net/vascan/search/

Evergreen Native Plant Database:
http://nativeplants.evergreen.ca/

Flora of North America, from the Flora of North America Association:
http://www.fna.org/FNA

Flora Ontario – Integrated Botanical Information System (FOIBIS), from the University of Guelph:
http://www.uoguelph.ca/foibis/

Native American Ethnobotany Database, from the University of Michigan-Dearborn:
http://herb.umd.umich.edu/

Natureserve:
http://natureserve.org/explorer/

United States Department of Agriculture Plants Database: Natural Resources Conservation Service:
http://plants.usda.gov/

Red swamp currant (*Ribes triste*)

Index to Common and Scientific Names

Entries in **boldface** type refer to the primary species accounts.

A

Acer spicatum, 133
Actaea spp., 202–203
  alba. See *A. pachypoda*
  arguta. See *A. rubra*
  eburnean. See *A. rubra*
  pachypoda, 203
  rubra, 203
alder
  black. *See* winterberry
  dwarf. *See* buckthorn,
    alder-leaved
Amelanchier spp., 98–101
  alnifolia, 101
  arborea, 100
  bartramiana, 101
  canadensis, 100
  humilis. See *A. alnifolia*
  laevis, 101
  sanguinea, 101
  spicata, 100
apples, 32–35
  common, 33–34
  Indian. *See* may-apple
  thorn. *See* hawthorn,
    black
  wild. *See* common a.
apricot, 35, 59
Aralia spp., 172–173
  hispida, 173
  nudicaulis, 173
  racemosa, 173
Arctostaphylos spp.,
    122–123
  alpina, 123
    var. *rubra,* 123
  uva-ursi, 123
Arctuous
  alpina. See *Arctostaphylos*
    alpina
  rubra. See *Arctostaphylos*
    uva-ursi
Arisaema triphyllum,
    206–207
Aronia melanocarpa. See
    Photinia melanocarpa
arrowwood, downy, 134

B

bake-apple. *See* cloudberry
baneberries, 202–203
  red, 202, 203
  western. *See* red b.

white, 202, 203
barberries, 82–83
  common, 83
  European. *See* common b.
  holly-leaved. *See* Oregon-
    grape, tall
  Japanese, 83
bearberries, 122–123
  alpine, 123
  black alpine. *See* alpine b.
  common, 123
  red, 123
Berberis spp., 82–83
  aquifolium. See *Mahonia*
    aquifolium
  thunbergii, 83
  vulgaris, 83
bilberry, bog. *See* blueberry,
  bog
bittersweet, European. *See*
  nightshade, bittersweet
blackberries, 66–69
  Allegheny. *See* common b.
  Canada. *See* smooth b.
  common, 68
  high-bush. *See* common b.
  smooth, 68
blackcurrant, swamp. *See*
  currant, prickly
blueberries, 110–115, 124
  bog, 113
  dwarf, 113
  high bush, 113
  low sweet. *See* low-bush b.
  low-bush, 113
  oval-leaved, 114
  velvet-leaf, 114
buckbrush. *See* snowberry,
  northern
buckthorns, 184–185. *See*
    glossy b.
  alder. *See* glossy b.
  alder-leaved, 185
  European, 185
  glossy, 185
buffaloberry, russet. *See*
  soapberry
bunchberry, 174–175
bush-cranberries, 132–137
  American, 133. *See*
    cranberry, European
    highbush

high, 112, 134. *See*
    cranberry, marsh
  low. *See* high b.
bush, fever. *See* winterberry

C

catberry. *See* holly, mountain
Caulophyllum spp., 210–
    211
  giganteum, 211
  thalictroides, 211
cedar, eastern red, 41
Celtis
  occidentalis, 142–143
  tenuifolia, 143
checkerberry. *See*
  wintergreen
cherries, 56–59
  bird. *See* pin c.
  dwarf. *See* sand c.
  fire. *See* pin c.
  ground. *See* ground-
    cherries
  low sand. *See* sand c.
  Pennsylvania. *See* pin c.
  pin, 57
  rum. *See* wild black c.
  sand, 57–58
  timber. *See* wild black c.
  wild. *See* chokecherry
  wild black, 58–59
  wine. *See* wild black c.
chokeberry, black, 54–55
chokecherry, 60–63
  black. *See* chokeberry,
    black. *See* cherry, wild
    black
Cimicifuga racemosa, 210
Clintonia borealis, 148–149
clintonia, yellow, 148–149
cloudberry, 68, 73, **76–77**
cohoshes
  black, 210
  blue, 210–211
  giant blue, 211
  purple-flowered blue. *See*
    giant blue c.
  yellow-flowered blue. *See*
    blue c.
Comandra livida. See
    Geocaulon lividum
comandra, northern,
    178–179

cornel, silky. *See* dogwood, silky
Cornus spp., 102–105, 174–175
   alternifolia, 103
   canadensis, 174–175
   florida, 104
   obliqua, 105
   rugosa, 104–105
   sericea, 104
   stolonifera. See *C. sericea*
   unalaschensis. See *C. canadensis*
cowberry. *See* lingonberry
crabapple, 32, 33
   eastern. *See* wild c.
   sweet. *See* wild c.
   wild, 34
crackleberry. *See* huckleberry, black
cranberries, 112, 116–119, 124
   bog, 118–119
   European highbush, 133, 134
   high bush, 112, 134. *See* marsh c.
   large, 119
   low bush. *See* high bush c. *See* lingonberry
   marsh, 136
   mountain. *See* lingonberry
   rock. *See* lingonberry
   small. *See* bog c.
Crataegus spp., 42–47
   allwangeriana. See *C. pedicellata*
   aulica. See *C. pedicellata*
   chrysocarpa, 46–47
   crus-galli, 44
   douglasii, 44
   flabellata, 46
   macrosperma, 46
   mollis, 45
   monogyna, 45
   pedicellata, 47
   punctata, 44–45
   succulenta, 47
creeper, thicket. *See* Virginia creeper, false
crowberry, black, 124–125
cucumber root, Indian, 162–163
currants, 88–91

American black. *See* wild black c.
   bristly black. *See* prickly c.
   golden, 88, 90
   Hudson Bay. *See* northern black c.
   northern black, 90
   prickly, 96–97
   red swamp, 90–91
   skunk, 88, 91
   wild black, 91
   wild red. *See* red swamp c.

D
deerberry, 109
Devil's club, 192–193
dewberries, 74–75. *See* raspberry, dwarf. *See* dewberry, swamp
   bristly. *See* swamp d.
   northern, 75
   swamp, 75
Dirca palustris, 214–215
Disporum trachycarpum. See *Prosartes trachycarpa*
dog-berry. *See* mountain ash, showy
dogberry. *See* gooseberry, prickly
dogwoods, 102–105
   alternate-leaved, 103
   blue. *See* alternate-leaved d.
   Canada. *See* bunchberry
   dwarf. *See* bunchberry
   eastern flowering, 104
   pagoda. *See* alternate-leaved d.
   red-osier, 102, 103, 104
   round-leaved, 104–105
   silky, 105
doll's eyes. *See* baneberry, white

E
Echinopanax horridum. See *Oplopanax horridus*
Elaeagnus commutata, 144–145
elder
   American. *See* elderberry, common
   dwarf. *See* sarsaparilla, bristly
   red-berried. *See* elderberry, red

elderberries, 128–131
   common, 130
   red, 130
Empetrum nigrum, 124–125

F
fairybells
   rough-fruited, 160–161
   wartberry. *See* fairybells, rough-fruited
false wintergreens, 120–121
   hairy, 121
false-hellebore, green, 151
Fragaria spp., 166–169
   chiloensis, 167–168
   vesca, 168
   virginiana, 168
Frangula alnus. See *Rhamnus frangula*

G
Gaultheria spp., 120–121
   hispidula, 121
   procumbens, 121
Gaylussacia baccata, 106, 108–109
Geocaulon lividum, 178–179
ginseng, 170–171
   American, 171
   dwarf, 171
gooseberries, 92–95
   bristly wild. *See* northern g.
   cape. *See* ground-cherry
   northern, 94
   prickly, 94
   smooth. *See* northern g.
   swamp. *See* currant, prickly
   wild, 95
grape
   fox, 126
   frost. *See* riverbank g.
   Oregon. *See* Oregon-grape, tall
   riverbank, 126–127
   summer, 126
   wild, 219
ground-cherries, 176–177
   clammy, 177
   large white. *See* white g.
   large-flowered. *See* white g.
   smooth, 177

Virginia. *See* smooth g.
white, 177

H

hackberry
American. *See* common h.
common, 142–143
dwarf, 143
hawthorns, 42–47
bigfruit, 46
black, 44
cockspur, 44
common, 45
dotted, 44–45
downy, 45
English. *See* common h.
fanleaf, 46
fireberry, 46–47
fleshy, 47
golden-fruited. *See* fireberry h.
long-spined. *See* fleshy h.
New England. *See* fanleaf h.
red. *See* downy h.
scarlet, 47
succulent. *See* fleshy h.
western. *See* black h.
hemlock, ground. *See* yew, Canadian
hobble bush. *See* hobbleberry
hobbleberry, 134–136
holly
Canada. *See* winterberry
mountain, 182–183
honeysuckles, 194–197
American fly. *See* fly h.
bracted. *See* twinberry, black
Canadian fly. *See* fly h.
fly, 196
glaucous. *See* twining h.
hairy, 196
limber. *See* twining h.
mountain fly, 196
northern. *See* twinberry, black
red. *See* twining h.
swamp fly, 197
tartarian, 197
twinflower. *See* twinberry, black
twining, 197

huckleberries, 106–109, 112
black, 106, 107, 108–109
black mountain, *See* black h.
squaw. *See* deerberry
thinleaf. *See* black h.
western. *See* blueberry, bog

I, J, K

Ilex
  mucronata. See *Nemopanthus mucronatus*
  verticillata, 180–181
jack-in-the-pulpit, 206–207
small. *See* jack-in-the-pulpit
juneberry. *See* serviceberry
downy. *See* serviceberry, downy
low, 100
mountain, 101
junipers, 36–41
common, 40
creeping, 40–41
ground. *See* common j.
Juniperus spp., 36–41
  communis, 40
  horizontalis, 40–41
  virginiana, 41
kinnikinnick. *See* bearberry, common

L

leatherwood, 214–215
eastern. *See* leatherwood
Leucophysalis spp., 176–177
  grandiflora, 177
lily-of-the-valley
feathery false. *See* Solomon's-seal, common false
starry false. *See* Solomon's-seal, star-flowered false
wild. *See* mayflower, Canada
lily, yellow blue-bead. *See* clintonia, yellow
lingonberry, 119
liverberry. *See* twisted-stalk, clasping

Lonicera spp., 194–197
  canadensis, 196
  dioica, 197
  hirsuta, 196
  involucrata, 195–196
  oblongifolia, 197
  tatarica, 197
  villosa, 196

M

Mahonia aquifolium, 84–85
Maianthemum spp., 154–159
  canadense, 154–155
  racemosum, 157
  stellatum, 158
  trifolium, 158
Malus spp., 32–35
  coronaria, 34
  pumila, 33–34
mandarin
rough-fruited. *See* fairybells, rough-fruited
white. *See* twisted-stalk, clasping
mandrake, wild. *See* may-apple
maple, mountain, 133
May tree. *See* hawthorn, common
may-apple, 212–213
mayflower, Canada, 154–155
Medeola virginiana, 162–163
Menispermum canadense, 127, 218–219
Mitchella repens, 146–147
moonseed, 127
  Canada, 218–219
mooseberry. *See* bush-cranberry, high
moosewood. *See* leatherwood
Morus spp., 80–81
  alba, 81
  rubra, 81
mountain ashes, 35, 48–49
American, 49
European, 49
northern. *See* showy m.
showy, 49
mulberries, 80–81
red, 81
white, 81

N

nagoonberry. *See* raspberry, Arctic dwarf
nannyberry, 136
Nemopanthus mucronatus, 182–183
nettletree. *See* hackberry, common
nightshades, 204–205
bitter. *See* bittersweet n.
bittersweet, 204, 205
climbing. *See* bittersweet n.
eastern black, 205
west Indian. *See* eastern black n.
Nudum cassinoides. See *Viburnum cassinoides*

O

Oplopanax horridus, 192–193
Oregon-grape, tall, 84–85
osier, green. *See* dogwood, alternate-leaved
Oxycoccus
macrocarpus. See *Vaccinium macrocarpon*
microcarpus. See *Vaccinium oxycoccos*
oxycoccos. See *Vaccinium oxycoccos*
quadripetalus. See *Vaccinium oxycoccos*

P

Panax spp., 170–171
ginseng, 170
quinquefolius, 171
trifolius, 171
Parthenocissus
quinquefolia, 216, 217
vitacea, 216–217
partridge berry, 146–147. *See* lingonberry
peach, 35, 59
pear, 35
pembina. *See* cranberry, European highbush
Photinia melanocarpa, 54–55
Physalis spp., 176–177
heterophylla, 177
virginiana, 177
Phytolacca americana, 208–209

plum, 59
Podophyllum peltatum, 212–213
poison-ivy, 188–189
eastern, 189
vining, 189
western, 189
poison-oak, 188–189
northern. *See* poison-ivy, western
pokeberry. *See* pokeweed
pokebush. *See* pokeweed
pokeweed, 208–209
American. *See* pokeweed
Polygonatum spp., 164–165
biflorum, 165
pubescens, 165
Prosartes trachycarpa, 160–161
Prunus spp., 56–63
pensylvanica, 57
pumila, 57–58
serotina, 58–59
virginiana, 60–63

R

raisin, northern wild, 136
raspberries, 68, 70–73
American red. *See* wild red r.
Arctic dwarf, 71
black, 71
dwarf, 71
eastern black. *See* black r.
flowering, 72
northern dwarf. *See* Arctic dwarf r.
trailing. *See* dwarf r.
wild red, 72
Rhamnus spp., 184–185
alnifolia, 185
cathartica, 185
frangula, 185
Rhus spp., 64–65
aromatica, 65
glabra, 65
radicans. See *Toxicodendron radicans*
var. *rydbergii.* See *Toxicodendron rydbergii*
typhina, 65
vernix. See *Toxicodendron vernix*

Ribes spp., 88–97
americanum, 91
aureum, 90
cynosbati, 94
glandulosum, 91
hirtellum, 95
hudsonianum, 90
lacustre, 96–97
oxyacanthoides, 94, 95
setosum. See *R. oxyacanthoides*
triste, 90–91
root
papoose. *See* cohosh, blue
squaw. *See* cohosh, blue
Rosa spp., 50–53
acicularis ssp. *sayi,* 51
blanda, 52
multiflora, 51
palustris, 52
rugosa, 51–52
roses
baby. *See* multiflora r.
multiflora, 51
prickly wild, 51
rugosa, 51–52
smooth wild, 52
swamp, 52
wild, 50–53
Rowan tree. *See* mountain ash, European
Rubus spp., 66–79
acaulis. See *R. arcticus*
allegheniensis, 68
arcticus, 71
canadensis, 68
chamaemorus, 76–77
flagellaris, 75
hispidus, 75
idaeus, 72
occidentalis, 71
odoratus, 72
parviflorus, 78–79
pubescens, 71
strigosis. See *R. idaeus*

S

salmonberry, 68, 73
Sambucus spp., 128–131
melanocarpa. See *S. racemosa*
nigra, 130
pubens. See *S. racemosa*
racemosa, 130

sarsaparillas, 172–173
  bristly, 173
  hairy. *See* bristly s.
  wild, 173
Saskatoon. *See* serviceberry
  berry, 101
sassafras, 86–87
Sassafras albidum, 86–87
serviceberries, 98–101. *See*
  Saskatoon berry
  Canada, 100. *See*
    serviceberry
  downy, 100
  dwarf. *See* juneberry, low
  red-twigged, 101
  round-leaved. *See* red-
    twigged s.
  running. *See* juneberry,
    low
  smooth, 101
shadbush. *See* serviceberry.
  See serviceberry, downy.
  See serviceberry, red-
  twigged
sheepberry. *See* nannyberry
Shepherdia canadensis,
  140–141
silverberry, 144–145
Smilacina
  racemosa. See
    Maianthemum
    racemosa
  stellata. See
    Maianthemum
    stellatum
  trifolia. See
    Maianthemum trifolium
smilacina, three-leaved.
  See Solomon's-seal, three-
  leaved
snake berry. *See* baneberry,
  red
snowberries, 198–201. *See*
  false-wintergreen, hairy
  common, 200
  creeping. *See* false-
    wintergreen, hairy
  northern, 200
soapberry, 140–141
Solanum spp., 204–205
  dulcumara, 205
  ptychanthum, 205
Solomon's-seals, 164–165
  common false, 157

false, 156–159
  great. *See* Solomon's-seal,
    smooth
  hairy, 165
  little false. *See* star-
    flowered false S.
  smooth, 165
  star-flowered false, 158
  three-leaved, 158
Sorbus spp., 48–49
  americana, 49
  aucuparia, 49
  decora, 49
spikenard, 173
  false. *See* Solomon's-seal,
    common false
squashberry. *See* bush
  cranberry, high
strawberries, 166–169
  coastal, 167–168
  common. *See* wild s.
  wild, 167–168
  woodland, 168
Streptopus spp., 150–153
  amplexifolius, 153
  lanceolatus, 153
  roseus. See *S. lanceolatus*
sumacs, 64–65
  fragrant, 65
  poison, 190–191
  smooth, 65
  staghorn, 65
Symphoricarpos spp.,
  198–201
  albus, 200
  occidentalis, 200

T

Taxus canadensis, 186–187
teaberry, eastern. *See*
  wintergreen
thimbleberry, 68, 73, **78–79**
thorn, cockspur. *See*
  hawthorn, cockspur
toadflax, false. *See*
  comandra, northern
tomato, beach. *See* rose,
  rugosa
Toxicodendron spp.,
  188–191
  radicans, 189
    spp. *negundo.* See
      Toxicodendron
      radicans
      var. *radicans*

    var. *radicans,* 189
    var. *rydbergi.* See
      Toxicodendron
      rydbergii
  rydbergii, 189
  vernix, 190–191
turnip, Indian. *See* jack-in-
  the-pulpit
twinberry, black, 195–196
twisted-stalks, 150–153
  clasping, 153
  clasping-leaved. *See*
    clasping t.
  rosy, 153
two-eyed berry. *See*
  partridge berry

V

Vaccinium spp., 106–119
  angustifolium, 114
  caespitosum, 113
  corymbosum, 113
  macrocarpon, 119
  membranaceum, 106, 107,
    109
  myrtilloides, 114
  occidentale. See *V.*
    uliginosum
  ovalifolium, 114
  oxycoccos, 118–119
  stamineum, 109
  uliginosum, 113
  vitis-idaea, 119
Viburnum spp., 132–137
  acerifolium, 136
  alnifolium, 134–136
  cassinoides, 136
  edule, 112, 134
  lantana, 138–139
  lentago, 136
  opulus, 134
  rafinesquianum, 134
  trilobum, 136
    var. *americanum.* See
      V. opulus
viburnum, mapleleaf, 136
Virginia creeper, 216, 217
  false, 216–217
Vitis spp., 219
  aestivalis, 126
  labrusca, 126
  riparia, 126–127

W

wayfaring tree, 138–139

whitehaw. *See* hawthorn,
    dotted
whortleberry. *See* bearberry,
    alpine
wicopy. *See* leatherwood
willow, wolf. *See* silverberry

winterberry, 180–181
    American. *See*
    winterberry
wintergreen, 121
withe rod. *See* raisin,
    northern wild
withered. *See* raisin,
    northern wild

wolfberry. *See* snowberry,
    northern
woodbine, grape. *See*
    Virginia creeper, false

Y, Z
yew, Canadian, 186–187
    western, 186
zereshk, 82

Photo & Illustration Credits

Photo and illustration numbers refer to page number and location letter on the page. Page location is identified by sequential letters (a, b, c, d, etc.) for each page, running top to bottom on the left column, then top to bottom on the right column. Photo and illustration location letters are identified separately.

Example Page

Photo Credits

Wasyl Bakowsky: 191b. **Lee Beavington:** 14, 18b, 19a, 23, 24, 26b, 28, 29, 36, 38a, 39a, 60, 61, 62, 63a, 63b, 78, 79, 84, 96, 97, 98, 102, 103, 105, 106, 110, 111, 114a, 114b, 115, 118, 119, 122, 128, 130, 131, 132, 135c, 150, 152a, 153a, 153b, 156, 158, 167a, 168a, 174, 175, 192, 193, 198, 199. **Frank Boas:** 93, 107. **Todd Boland:** 17, 44, 46, 59a, 67b, 68, 83, 94, 101, 148, 149a, 149b, 154, 155, 159c, 183a, 183b, 183c, 196. **Charles T. Bryson/USDA Agricultural Research Service/Bugwood.org (CC):** 163. **Will Cook:** 214. **Alain Cuerrier:** 41b, 180, 182, 203, 208. **Paul S. Drobot:** 164, 165. **Jerry Drown:** 212. **David Duhl:** 170. **Mary Gartshore:** 81. **Cory Harris:** 32, 33a, 33b, 35, 53a, 53b, 54, 64, 82, 117, 126, 127, 133a, 137, 146, 159b, 184, 191a, 204, 205, 206, 209, 216, 219. **Neil Jennings:** 18c, 144, 145a. **Krista Kagume:** 189. **Linda Kershaw:** 19b, 20, 27, 30, 39b, 40, 41a, 45a, 45b, 48, 56, 57, 65, 67a, 72, 88, 89, 95b, 123, 125, 129, 133b, 138, 142, 143, 147b, 151, 152b, 157, 160, 161, 167b, 168b, 172, 179b, 179c, 179d, 181, 190, 201b, 201c, 202, 213, 225. **kirybabe/Flickr.com (CC):** 75. **Liz Klose:** 147a. **Ron Long:** 145b. **Glen Lumis:** 104b, 139a, 139b. **Robin Marles:** 104a, 159a, 171, 201a, 218. **Tim Matheson:** 34. **James H. Miller & Ted Bodner/Southern Weed Science Society/Bugwood.org (CC):** 18a, 74. **Walter Muma:** 66, 176, 210, 211. **pellaea/Jason Hollinger/Flickr.com (CC):** 162. **Dave Powell/USDA Forest Service/Bugwood.org (CC):** 177. **Richtid/Flickr.com (CC):** 215. **Virginia Skilton:** 16, 42, 92, 95a, 188, 194. **Superior National Forest:** 179a. **Thinkstock 2010:** 1. **Robert D. Turner and Nancy J. Turner:** 11, 12, 31, 37, 38b, 50, 52, 70, 73a, 73b, 76, 77, 90, 112, 116, 120, 124, 135a, 135b, 140, 141, 166, 217. **Per Verdonk:** 26a, 55, 59b, 80, 87, 100, 186.

Images above marked "(CC)" are available for reuse under the Creative Commons Attribution - Share Alike License (or variants), which generally means you are free: to Share—to copy, distribute and transmit the work; and to Remix—to adapt the work under the following conditions: you must attribute the work in the manner specified by the author or licensor (but not in any way that suggests that they endorse you or your use of the work); and if you alter, transform, or build upon this work, you may distribute the resulting work only under the same, similar or a compatible license. Please see actual license for details.

Illustration Credits

Frank Burman: 37a, 37b, 40a, 43, 44a, 47a, 48, 51a, 51b, 52, 53, 56, 58b, 58c, 65a, 67, 71a, 71b, 72a, 72b, 77a, 77b, 82, 85, 88, 90, 91a, 91b, 91c, 94, 109b, 111, 112, 113a, 113b, 114, 118a, 118b, 119, 121b, 123a, 123c, 125, 127, 131a, 135, 136b, 137a, 137b, 147, 152a, 155a, 155b, 157, 158a, 159, 161a, 164, 168b, 169, 171, 172, 173a, 173b, 181a, 181b, 183, 185a, 185b, 187, 188, 189a, 189b, 191a, 191b, 193, 197a, 197b, 200, 205, 207a, 209, 213. **Linda Kershaw:** 13, 14, 15a, 15b, 15c, 15d, 46a, 47c, 129b, 220, 222. **George Penetrante:** 33, 34, 55, 75a, 75b, 81, 86, 109a, 121a, 139, 143b, 149a, 149b, 158c, 163a, 163b, 165a, 165b, 177, 178, 203, 207b, 211, 215, 217, 219. **Ian Sheldon:** 35, 38, 40b, 44b, 45, 46b, 46c, 47b, 49a, 49b, 58a, 59, 61, 62a, 62b, 65b, 79a, 79b, 80, 87, 97a, 97b, 99, 100a, 100b, 101, 103, 105a, 105b, 106, 107, 123b, 129a, 131b, 136a, 138, 141, 143a, 145a, 145b, 152b, 153, 158b, 161b, 161c, 168a, 175, 195a, 195b, 199a, 199b, 202.

About the Authors

FIONA HAMERSLEY CHAMBERS was born in Vancouver. She holds an undergraduate degree from the University of Victoria in French and Environmental Studies (1994), a Masters of Science in Environmental Change and Management from Oxford University (1998), and a Masters in Environmental Design from the University of Calgary (1999). Speaking English, French and Spanish, she has travelled extensively throughout Europe, Australia, New Zealand, Mexico and Central America and has a strong interest in learning about traditional plant uses wherever she goes. Fiona has taught Environmental Studies at the University of Victoria since 1999, and ethnobotany at Pacific Rim College since 2009. She currently divides her time between teaching, running a small organic farm and food plant nursery (www.metchosinfarm.ca), writing books and academic papers, and raising two energetic boys who also love plants, animals and bugs.

CORY HARRIS was born in Toronto and holds a PhD in Biology from the University of Ottawa where he studied medicinal plants of the Canadian boreal forest. Currently a Banting Postdoctoral Fellow at McGill University's School of Dietetics and Human Nutrition and the Centre for Indigenous Peoples' Nutrition and Environment, Cory studies the roles of plant foods and medicines in health—from nutrition and pharmacology to community and environment—with a special interest in berries. He has taught nutrition and plant physiology at the university level and tries to spread his love of wild harvesting with his family and friends. In August and September, you can often find him with red- or purple-stained fingers and thorn-torn pants (and ankles).